'Henry Gee's whistle-stop account of the story of life (and death – lots of death) on Earth is both fun and informative. Even better, it goes beyond the natural human inclination to see ourselves as special and puts us in our proper place in the cosmic scheme of things.' John Gribbin

'This is now the best book available about the huge changes in our planet and its living creatures, over the billions of years of the Earth's existence. Continents have merged and broken up; massive volcanic eruptions have repeatedly reset the clock of evolution; temperatures, atmospheric gases, and sea levels have undergone big swings; and new ways of life have evolved. Henry Gee makes this kaleidoscopically changing canvas of life understandable and exciting. Who will enjoy reading this book? – Everybody!'

Jared Diamond, Pulitzer Prize-winning author of *Guns, Germs, and Steel*, Distinguished Professor of Geography, Physiology and Public Health, UCLA

'Don't miss this delightful, concise, sweeping masterpiece! Gee brilliantly condenses the entire, improbable, astonishing history of life on earth – all 5 billion years – into a charming, zippy and scientifically accurate yarn. I honestly couldn't put this book down, and you won't either.'

Daniel E. Lieberman, Edwin M. Lerner II Professor of Biological Sciences, Harvard University

A (VERY) SHORT HISTORY OF LIFE ON EARTH

Henry Gee is a senior editor at *Nature* and the author of several books, including *Jacob's Ladder*, *In Search of Deep Time*, *The Science of Middle-earth*, and *The Accidental Species*. He has appeared on BBC television and radio and NPR's *All Things Considered*, and has written for *The Guardian*, *The Times*, and *BBC Focus*. He lives in Cromer, Norfolk, England, with his family and numerous pets.

Also by Henry Gee

NON-FICTION

The Letts Pocket Guide to Fossils

Before The Backbone: Views on the Origin of the Vertebrates

Deep Time: Cladistics, the Revolution in Evolution

A Field Guide to Dinosaurs: The Essential Handbook for Travellers in the Mesozoic

Jacob's Ladder: The History of the Human Genome

The Science of Middle-earth

The Accidental Species: Misunderstandings of Human Evolution

Across The Bridge: Understanding the Origin of the Vertebrates

FICTION

The Sigil Trilogy

By The Sea

Hunting Unicorns and Other Stories

EDITED COLLECTIONS

Shaking the Tree: Readings from Nature in the History of Life

Rise of the Dragon: Readings from Nature on the Chinese Fossil Record

Futures from Nature: 100 Speculative Fictions from the Pages of the Leading Science Journal

Futures 2: Science Fiction from the Leading Science Journal

FOR YOUNGER READERS

Defiant the Guinea-Pig — Firefighter!

'This history of life on Earth exhilaratingly whizzes through billions of years . . . Gee is a marvellously engaging writer, juggling humour, precision, polemic and poetry to [illegible]ch his impossibly telescoped account . . . Dizzying and [illegible]rating . . . To weave such interconnected wonders int[illegible] book the size of a modest novel is essentially an [illegible]se in precis and a bravura demonstration of the edito[illegible]s art . . . Gee's final masterstroke is to make human sens[illegible] and real tragedy, from his . . . story's glaring spoiler: t[illegible]fe dies at the end.' *The Times*

'[illegible]s] lively, lyrical history covers 4.6 billion years, from [illegible]ia through dinosaurs to mammals including Homo [illegible]s. Humans, Gee says, will eventually become a thin [illegible]n sedimentary rock, to be eroded as dust that sinks [illegible] ocean bottom.' *Nature*

'[illegible]sive and satisfying . . . Each of life's phases and each [illegible]f the evolutionary ladder [is] described both poetically [illegible]ith a satisfying sense of order . . . Gee dissolves life's [illegible]boggling complexity into something digestible for [illegible]one . . . He plunges us back in time but also casts us [illegible] to a juvenile state of wonder. If you're prone to fleeting [illegible]ents in the midst of daily tasks in which you stop to [illegible]tion how all this precious life came to be, the answers [illegible]be found conveniently packed within these pages.'

National Geographic

'This is the enlightening story of survival, illuminating the delicate balance in which life exists. Henry Gee's lyrical prose personifies creatures and conveys life's evolutionary steps with alluring intimacy.' *The Telegraph*

'Assuming the role of a peripatetic tour guide, Henry Gee . . . takes the reader on an exuberant romp through evolution, like a modern-day Willy Wonka of genetic space. Gee's grand tour enthusiastically details the narrative underlying life's erratic and often whimsical exploration of biological form and function . . . Gee has . . . succeeded in producing a seamless and highly compressed account of life's grand narrative, spanning its full duration of about 4.6 billion years. It is a tale of resilience and tenacity, and his writing is evocative and filled with humor.' *The Washington Post*

'This one is easy to sum up. Brilliant book. Buy it.'
Popular Science

'"Once upon a time . . ." The opening words of Henry Gee's new book give notice that what follows will be a story – and a dazzling, beguiling story it is, told at an exhilarating pace . . . [a] hugely enjoyable page-turner' *Literary Review*

'Gee's prose is so infectiously enthusiastic, and his tone so accessible, that you'll find yourself racing through as if you were reading a novel – and you'll never find yourself scrambling for a good fact to wheel out at an awkward pause in conversation again.' *Reader's Digest*

A (VERY) SHORT HISTORY OF LIFE ON EARTH

4.6 Billion Years
in 12 Chapters

HENRY GEE

PICADOR

First published 2021 by Picador

This edition published 2022 by Picador
an imprint of Pan Macmillan
The Smithson, 6 Briset Street, London EC1M 5NR
EU representative: Macmillan Publishers Ireland Ltd, 1st Floor,
The Liffey Trust Centre, 117–126 Sheriff Street Upper, Dublin 1, D01 YC43
Associated companies throughout the world
www.panmacmillan.com

ISBN 978-1-5290-6058-4

1 3 5 7 9 8 6 4 2

A CIP catalogue record for this book is available from the British Library.

Typeset in Janson Text LT Std by
Palimpsest Book Production Ltd, Falkirk, Stirlingshire

Printed and bound by CPI Group (UK) Ltd, Croydon, CR0 4YY

Visit www.picador.com to read more about all our books
and to buy them. You will also find features, author interviews and
news of any author events, and you can sign up for e-newsletters
so that you're always first to hear about our new releases.

To the memory of Jenny Clack (1947–2020)
Mentor, Friend

Contents

Timeline 1. Earth in the Universe

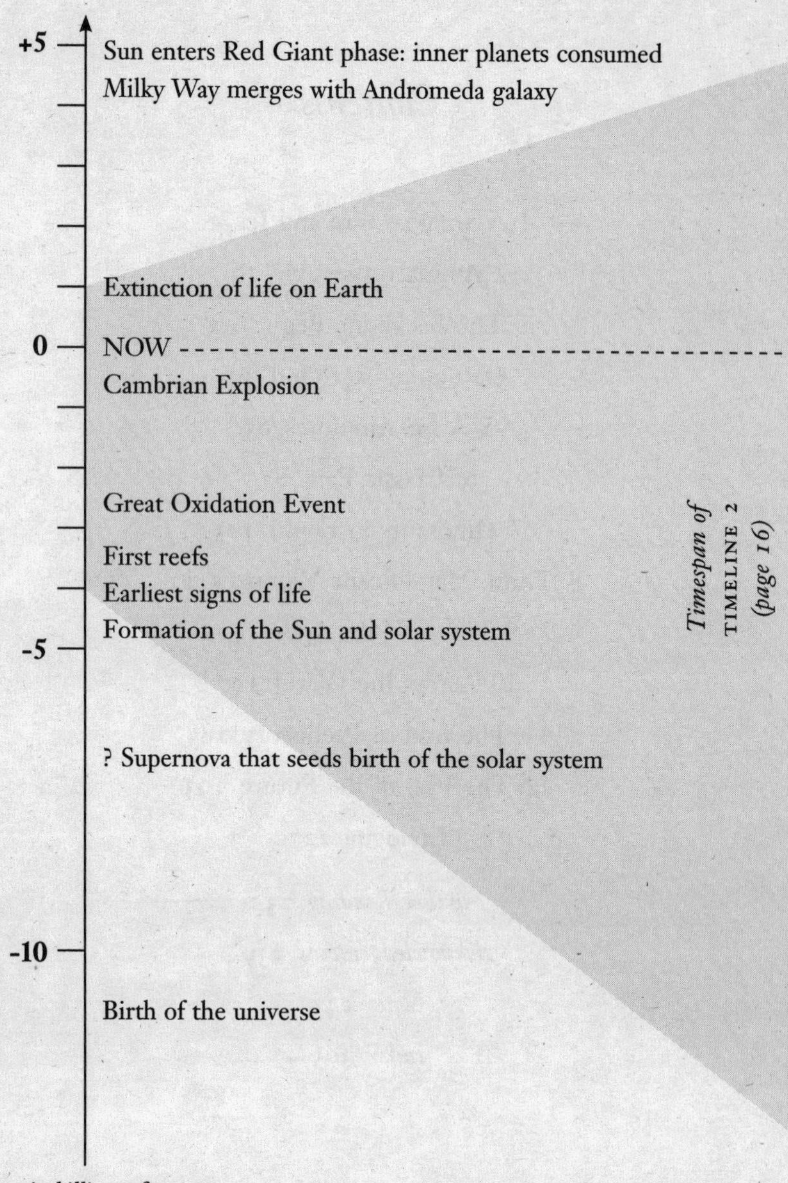
+5
Sun enters Red Giant phase: inner planets consumed
Milky Way merges with Andromeda galaxy
Extinction of life on Earth
0
NOW
Cambrian Explosion
Great Oxidation Event
First reefs
Earliest signs of life
Formation of the Sun and solar system
-5
? Supernova that seeds birth of the solar system
-10
Birth of the universe
Timespan of TIMELINE 2 (page 16)
Ages in billions of years before (minus) or after (plus) the present

I

A SONG *of* FIRE *and* ICE

Once upon a time, a giant star was dying. It had been burning for millions of years; now the fusion furnace at its core had no more fuel to burn. The star created the energy it needed to shine by fusing hydrogen atoms to make helium. The energy produced by the fusion did more than make the star shine. It was vital to counteract the inward pull of the star's own gravity. When the supply of available hydrogen began to run low, the star began to fuse helium into atoms of heavier elements such as carbon and oxygen. By then, though, the star was running out of things to burn.

The day came when the fuel ran out completely. Gravity won the battle: the star imploded. After the millions of years of burning, the collapse took a split second. It prompted a rebound so explosive that it lit up the universe – a supernova. Any life that might have existed in the star's own planetary system would have been obliterated. But in the cataclysm of its death were born the seeds of something new. Even heavier chemical elements, forged in the final moments of the star's life – silicon, nickel, sulphur and iron – were spread far and wide by the explosion.

Millions of years later, the gravitational shock wave of the supernova explosion passed through a cloud of gas, dust and ice. The stretch and squeeze of the gravitational wave made the cloud fall in on itself. As it contracted, it started to rotate. The pull of gravity squeezed the gas at the cloud's centre so much that atoms began to fuse together. Hydrogen atoms were pressed together, forming helium, creating light and heat. The circle of stellar life was complete. From the death of an ancient star emerged another, fresh and new – our Sun.

The cloud of gas, dust and ice was enriched with the elements created in the supernova. Swirling around the new Sun, it also coagulated into a system of planets. One of them was our Earth. The infant Earth was very different from the one we know today. The atmosphere would have been to us an unbreathable fog of methane, carbon dioxide, water vapour and hydrogen. The surface was an ocean of molten lava, perpetually stirred up by the impacts of asteroids, comets and even other planets. One of these was Theia, a planet about the same size as today's Mars.[1] Theia struck the Earth a glancing blow, and disintegrated. The collision blasted much of the Earth's surface into space. For a few million years, our planet had rings, like Saturn. Eventually the rings coalesced to create another new world – the Moon.[2] All this happened approximately 4,600,000,000 (4.6 billion) years ago.

Millions more years passed. The day came when the Earth had cooled enough for the water vapour in the atmosphere to condense and fall as rain. It rained for millions of years, long enough to create the first oceans. And oceans were all there were – there was no land. The Earth, once a ball of fire, had become a world of water. Not that things were any calmer. In those days the Earth spun faster on its axis than it does today. The new Moon loomed close above the black horizon. Each incoming tide was a tsunami.

A planet is more than a jumble of rocks. Any planet more than a few hundred kilometres in diameter settles out into layers over time. Less dense materials such as aluminium, silicon and oxygen combine into a light froth of rocks near the surface. Denser materials such as nickel and iron sink to the core. Today, the Earth's core is a rotating ball of liquid metal. The core is kept hot by gravity, and the decay of heavy radioactive elements such as uranium, forged in the final moments of the ancient supernova. Because the Earth spins, a magnetic field is generated in the core. The tendrils of this magnetic field reach right through the Earth and stretch far out into space. The magnetic field shields the Earth from the solar wind, a constant storm of energetic particles streaming from the Sun. These particles are electrically charged, and, repelled by the Earth's magnetic field, bounce off, or flow around the Earth and into space.

The Earth's heat, radiating outwards from the molten core, keeps the planet forever on the boil, just like a pan of water simmering on a stove. Heat rising to the surface softens the overlying layers, breaking up the less dense but more solid crust into pieces, and, forcing them apart, creates new oceans between. These pieces, the tectonic plates, are forever in motion. They bump against, slide past or burrow beneath one another. This movement carves deep trenches in the ocean floor and raises mountains high above it. It causes earthquakes and volcanic eruptions. It builds new land.

As the bare mountains were thrust skywards, vast quantities of the crust were sucked back into the depths of the Earth in deep ocean trenches, at the edges of the tectonic plates. Laden with sediment and water, this crust was drawn deep into the Earth's interior – only to return to the surface, changed into new forms. The ocean-floor sludge at the fringes of vanished continents might, after hundreds of millions of years, re-emerge in volcanic eruptions,[3] or be transformed into diamonds.

Amid all this tumult and disaster, life began. It was the tumult and disaster that fed it, nurtured it, made it develop and grow. Life evolved in the deepest depths of the ocean, where the edges of tectonic plates plunged into the crust; and where boiling hot jets of water, rich in minerals and under extreme pressure, gushed out from cracks in the ocean floor.

The earliest living things were no more than scummy membranes across microscopic gaps in rocks. They formed when the rising currents became turbulent and diverted into eddies, and, losing energy, dumped their cargo of mineral-rich debris[4] into gaps and pores in the rock. These membranes were imperfect, sieve-like, and, like sieves, allowed some substances to cross but not others. Even though they were porous, the environment inside the membranes became different from the raging maelstrom beyond, calmer, more ordered. A log cabin with a roof and walls is still a haven from the arctic blast outside, even if its door bangs and its windows rattle. The membranes made a virtue of their leakiness, using holes as gateways for energy and nutrients, and as exit points for wastes.[5]

Protected from the chemical clamour of the outside world, these tiny pools were havens of order. Slowly, they refined the generation of energy, using it to bud off small bubbles, each encased in its own portion of the parent membrane. This was haphazard at first, but gradually became more predictable, as a result of the development of an internal chemical template that could be copied and passed down to new generations of membrane-bound bubbles. This ensured that new generations of bubbles were, more or less, faithful copies of their parents. The more efficient bubbles began to thrive at the expense of those less well-ordered.

These simple bubbles found themselves at the very gates of life, in that they found a way to halt – if temporarily, and with great effort – the otherwise inexorable increase in entropy, the net amount of disorder in the Universe. Such is an essential property of life. These foamy lathers of soap-bubble cells stood as tiny clenched fists, defiant against the lifeless world.[6]

Perhaps the most amazing thing about life – apart from its very existence – is how quickly it began. It stirred itself into existence a mere 100 million years after the planet itself formed, in volcanic depths when the young Earth was still being bombarded from space by bodies large enough to create the major impact craters on the Moon.[7] By 3.7 billion years ago, life had spread from the permanent dark of the ocean depths to the sunlit surface waters.[8]

By 3.4 billion years ago, living things had started to throng together in their trillions to create reefs, structures visible from space.[9] Life on Earth had fully arrived.

These reefs were not composed of corals, however – they still lay almost 3 billion years into the Earth's future. They consisted of greenish, hair-thin threads and scuts of slime made from microscopic organisms called cyanobacteria – the same creatures that form the bluish-green scum on ponds today. They spread in sheets over rocks and lawns on the seabed, only to be buried by sand in the next storm: but conquering again, and being buried once again, building cushion-like mounds of layered slime and sediment. These mound-shaped masses, known as stromatolites, were to become the most successful and enduring form of life ever to have existed on this planet, the undisputed rulers of the world for 3 billion years.[10]

Life began in a world that was warm[11] but soundless apart from the wind and the sea. The wind stirred an air almost entirely free from oxygen. With no protective ozone layer in the upper atmosphere, the Sun's ultraviolet rays sterilized everything above the surface of the sea, or anything less than a few centimetres beneath the surface. As a means of defence, the cyanobacterial colonies evolved pigments to absorb these harmful rays. Once their energy had been absorbed, it could be put to work. The cyanobacteria used it to drive chemical reactions. Some of these fused carbon, hydrogen and oxygen atoms together to create sugars and starch. This is the process we call 'photosynthesis'. Harm had become harvest.

In plants today, the energy-harvesting pigment is called chlorophyll. Solar energy is used to split water into its constituent hydrogen and oxygen, releasing more energy to drive further chemical reactions. In the earliest days of the Earth, however, the raw materials were just as likely to have been minerals containing iron or sulphur. The best, however, was and remains the most abundant – water. But there was a catch. The photosynthesis of water produces as a waste product a colourless, odourless gas that burns anything it touches. This gas is one of the deadliest substances in the universe. Its name? Free oxygen, or O_2.

To the earliest life, which had evolved in an ocean and beneath an atmosphere essentially without free oxygen, it spelled environmental catastrophe. To set the matter into perspective, however, when cyanobacteria were making their first essays into oxygenic photosynthesis – 3 billion years ago, or more – there was rarely enough free oxygen at any time to count as more than a minor trace pollutant. But oxygen is so potent a force that even a trace spelled disaster to life that had evolved in its absence. These whiffs of oxygen caused the first of many mass extinctions in the Earth's history, as generation upon generation of living things were burned alive.

Free oxygen became more abundant during the Great Oxidation Event, a turbulent period between about 2.4 and 2.1 billion years ago, when, for reasons still unclear, the concentration of oxygen in the atmosphere at first rose sharply, to greater than today's value of 21 per cent, before settling down to a little below 2 per cent. Although still unbreathably tiny by modern standards, this had an immense effect on the ecosystem.[12]

An upsurge in tectonic activity buried vast quantities of carbon-rich organic detritus – the corpses of generation on generation of living things – beneath the ocean floor. This kept it away from oxygen's reach. The result was a surplus of free oxygen that could react with anything it touched. Oxygen etched the very rocks, turning iron to rust, and carbon to limestone.

At the same time, gases such as methane and carbon dioxide were scrubbed from the air, absorbed by the abundance of newly formed rock. Methane and carbon dioxide are two of the gases in the downy filling of the insulating blanket that keeps the Earth warm. They promote what we call the 'greenhouse effect'. Without them, the Earth plunged into the first and greatest of its many ice ages. Glaciers spread from pole to pole, covering the entire planet in ice for 300 million years. And yet the Great Oxidation Event and subsequent 'Snowball Earth' episode were the kinds of apocalyptic disasters in which life on Earth has always thrived. Many living things died, but life was spurred on to undergo its next revolution.

For the first 2 billion years in the Earth's story, the most sophisticated form of life was built on the bacterial cell. Bacterial cells are very simple, whether single, or glued together in sheets across the ocean floor, or in the long, angel-hair filaments of cyanobacteria. Each one, on its own, is tiny. As many bacteria could fit on the head of a pin as there were revellers who went to Woodstock, and with room to spare.[13]

Under a microscope, bacterial cells appear simple and featureless. This simplicity is deceptive. In terms of their habits and habitats, bacteria are highly adaptable. They can live almost anywhere. The number of bacterial cells in (and on) a human body is very much greater than the number of human cells in that same body. Despite the fact that some bacteria cause serious disease, we could not survive without the help of the bacteria that live in our guts and enable us to digest our food.

And the human interior, despite its wide variation in acidity and temperature, is, in bacterial terms, a gentle place. There are bacteria for which the temperature of a boiling kettle is as a balmy spring day. There are bacteria that thrive on crude oil; on solvents that cause cancer in humans; or even in nuclear waste. There are bacteria that can survive the vacuum of space; violent extremes of temperature or pressure; and entombment inside grains of salt – and do so for millions of years.[14]

Bacterial cells may be small, but they are famously gregarious. Different species of bacteria swarm together to trade chemicals. The waste products of one species might make a meal for another. Stromatolites – as we have seen, the first visible signs of life on Earth – were colonies of different kinds of bacteria. Bacteria can even swap portions of their own genes with one another. It is this easy trade that means, today, that bacteria can evolve resistance to antibiotics. If a bacterium doesn't have a resistance gene for a particular antibiotic, it can pick it up from the genetic free-for-all of other species with which it shares its environment.

It was the tendency of bacteria to form communities of different species that led to the next great evolutionary innovation. Bacteria took group living to the next level – the nucleated cell.

At some point before 2 billion years ago, small colonies of bacteria began to adopt the habit of living inside a common membrane.[15] It began when a small bacterial cell, called an archaeon,[16] found itself dependent on some of the cells around it for vital nutrients. This tiny cell extended tendrils towards its neighbours so they could swap genes and materials more easily. The participants in what had been a freewheeling commune of cells became more and more interdependent.

Each member concentrated only on one particular aspect of life.

Cyanobacteria specialized in harvesting sunlight, and became chloroplasts – the bright green specks now found in plant cells. Other kinds of bacteria devoted themselves to releasing energy from food, and became the tiny pink power-packs called mitochondria which are found in almost all cells that have nuclei, whether plant or animal.[17] Whatever their specialism, they all pooled their genetic resources in the central archaeon. This became the nucleus of the cell – the cell's library, repository of genetic information, its memory, and its heritage.[18]

This division of labour made life for the colony much more efficient and streamlined. What was once a loose colony became an integrated entity, a new order of life – the nucleated or 'eukaryotic' cell. Organisms made of eukaryotic cells – whether singly – unicellular – or lots together – multicellular – are called 'eukaryotes'.[19]

The evolution of the nucleus allowed for a more organized system of reproduction. Bacterial cells generally reproduce by dividing in half to create two identical copies of the parent cell. Variation from the addition of extra genetic material is piecemeal and haphazard.

In eukaryotes, by contrast, each parent produces specialized reproductive cells as vehicles for a highly choreographed exchange of genetic material. Genes from both parents are mixed together to create the blueprint for a new and distinct individual, different from either parent. We call this elegant exchange of genetic material 'sex'.[20] The increase in genetic variation as a consequence of sex drove an uptick in diversity. The result was the evolution of a wealth of different kinds of eukaryotes, and, over time, the emergence of gatherings of eukaryote cells to make multicellular organisms.[21]

Eukaryotes emerged, quietly and modestly, between around 1,850 and 850 million years ago.[22] They started to diversify around 1,200 million years ago, into forms recognizable as early single-celled relatives of algae and fungi, and unicellular protists, or what we used to call 'protozoa'.[23] For the first time, they ventured away from the sea and colonized freshwater ponds and streams inland.[24] Crusts of algae, fungi and lichens[25] began to adorn seashores once bare of life.

Some even experimented with multicellular life, such as the 1,200-million-year-old seaweed *Bangiomorpha*,[26] and the approximately 900-million-year-old fungus *Ourasphaira*.[27]

But there were stranger things. The earliest known signs of multicellular life are 2,100 million years old. Some of these creatures are as large as 12 centimetres across, so hardly microscopic, but they are so strange in form, to our modern eyes, that their relationship with algae, fungi or other organisms is obscure.[28] They could have been some form of colonial bacteria, but we cannot discount the possibility that there once lived entire categories of living organism – bacterial, eukaryote or something entirely other – that died out without leaving any descendants, and which we should therefore find hard to comprehend.

The first rumbles of an oncoming storm came from the rifting and break-up of a supercontinent, Rodinia. This included every significant landmass at the time.[29] One consequence of the break-up was a series of ice ages the like of which had not been seen since the Great Oxidation Event. They lasted 80 million years, and, like the earlier episode, covered the entire globe. But life responded once again by rising to the challenge.

Life entered the lists as a range of peaceable seaweeds, algae, fungi and lichens.

It emerged tough, mobile, and looking for trouble.

For if life on Earth was forged in fire, it was hardened in ice.

Timeline 2. Life on Earth

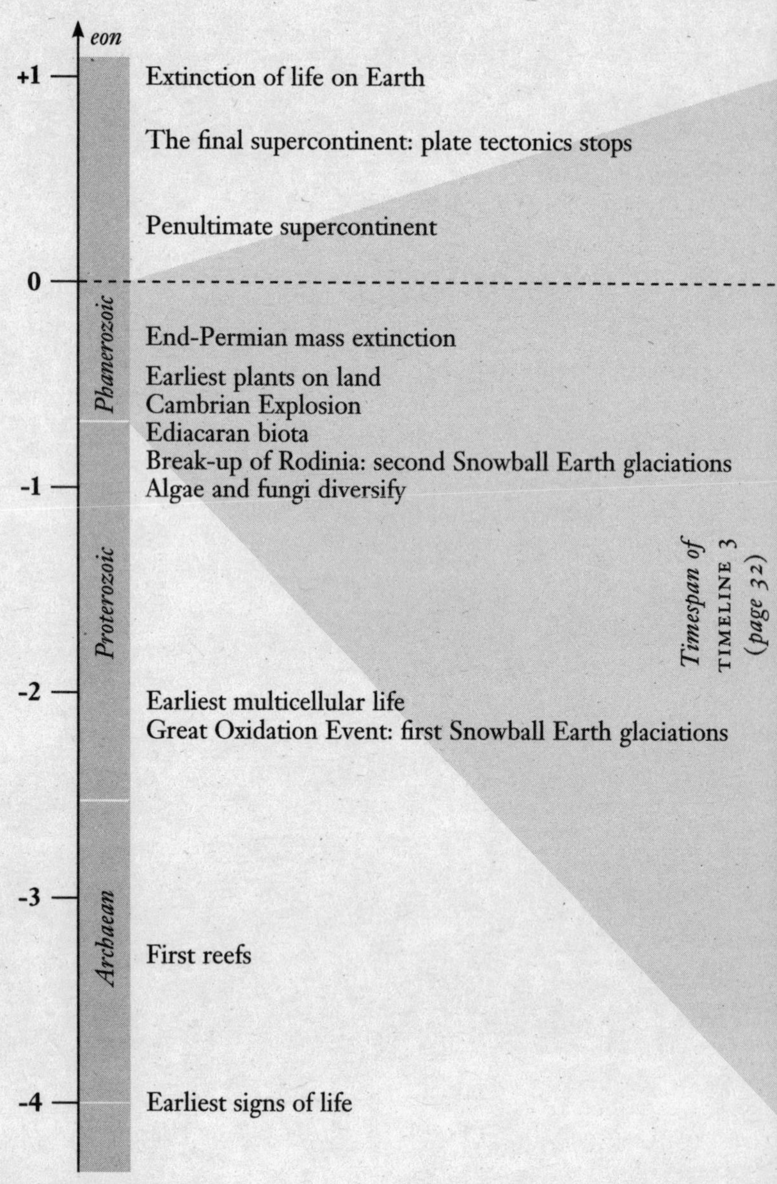
eon
+1
Extinction of life on Earth
The final supercontinent: plate tectonics stops
Penultimate supercontinent
0
Phanerozoic
End-Permian mass extinction
Earliest plants on land
Cambrian Explosion
Ediacaran biota
Break-up of Rodinia: second Snowball Earth glaciations
-1
Algae and fungi diversify
Proterozoic
Timespan of TIMELINE 3 (page 32)
-2
Earliest multicellular life
Great Oxidation Event: first Snowball Earth glaciations
-3
Archaean
First reefs
-4
Earliest signs of life
Ages in billions of years before (minus) or after (plus) the present

2

ANIMALS ASSEMBLE

The break-up of the supercontinent Rodinia began around 825 million years ago. It continued for almost 100 million years, leaving a ring of continents around the Equator. The break-up was accompanied by massive volcanic eruptions that brought vast amounts of volcanic rock to the surface, much of it the igneous rock called basalt. Basalt is easily weathered by rain and storm and many of the newly rifted landmasses were in the tropics, where greater heat and humidity makes weathering especially intense.

Wind and weather not only sloughed basalt into the oceans. They also tipped immense amounts of carbon-containing sediment into the depths, out of reach of oxygen. When carbon can be oxidized to form carbon dioxide, the Earth is warmed by the greenhouse effect. But with carbon removed from the atmosphere, the greenhouse effect stalls, and the Earth cools down. This dance of carbon, oxygen and carbon dioxide was to tap out a rhythm in the subsequent history of the Earth and the life that crawled on its face.

The result of the weathering of the fragments of Rodinia was that from around 715 million years ago, the Earth was pitched into a series of world-spanning ice ages that lasted around 80 million years.

As during the episode that followed the Great Oxidation Event more than a billion years earlier, these ice ages were spurs to evolution. They set the stage for the emergence of a new, more active kind of eukaryote – the animals.[1]

The carbon that washed into the sea entered an ocean which, apart from a thin layer close to the surface in contact with the atmosphere, contained almost no oxygen. Even so, the concentration of oxygen in the atmosphere was no more than a tenth of the present-day value, and even less in the sunlit ocean surface. This was too small to sustain any animal much larger than the full stop at the end of this sentence.

There were some animals, however, that managed to subsist on minute quantities of oxygen. These were the sponges. Sponges first appeared around 800 million years ago,[2] as Rodinia was starting to be torn apart.

Sponges were and are very simple animals. Although sponge larvae are small and mobile, adult sponges remain in one place their whole lives. An adult sponge is simply made, being no more than a shapeless mass of cells perforated by thousands of tiny holes, channels and spaces. The cells that line these spaces draw currents of water through them by beating hair-like extensions called cilia. Other cells absorb detritus from the water current. Sponges have no distinct organs or tissues. A live sponge pushed through a sieve and back into the water will pull itself together into a different shape, but one just as alive, just as functional. It is a simple life that requires little energy – and little oxygen.

But there is no reason to disparage that which is simple.

After the earliest sponges were settled, they changed the world.

Sponges living among the carpets of slime that draped the sea floor sieved particles of matter from the water. The volume of water pulled through one sponge in a day was small, but billions of sponges over tens of millions of years had an immense impact. The slow, steady work of sponges led to an even greater accumulation of sea-floor carbon, unavailable for reaction with oxygen. Sponges also cleared the water around them of detritus that would otherwise have been digested by oxygen-sucking decay bacteria. The result was a slow increase in the amount of oxygen dissolved in the sea, and in the air immediately above it.[3]

Far above the sponges, jellyfishes and small worm-like animals consumed smaller eukaryotes and bacteria in the plankton – the sunny region of the sea closest to the surface.[4] There was more oxygen in the surface waters to start with, but the carbon-rich bodies of creatures in the plankton, once dead, sank more quickly to the bottom, rather than remaining suspended in the water, removing more carbon from the reach of molecular oxygen. Which in turn left more oxygen to accumulate in the ocean and the atmosphere.

Although large enough that some would have been visible to human eyes without a microscope, many of the creatures that made up the plankton were small enough that nutrients and wastes could simply diffuse in and out of their bodies. Those that were a little larger evolved a particular place for nutrients to enter, and for waste to diffuse out again. That place was the mouth, although it also served double duty as an anus.

The development of a distinct anus in some species of otherwise undistinguished worms led to a revolution in the biosphere. For the first time, waste was concentrated into solid pellets, rather than being a general wash of dissolved excrement. These faeces sank quickly to the sea floor rather than diffusing slowly, and this led to a literal race to the bottom. Oxygen-sucking agents of decay began to concentrate their efforts near the sea floor, rather than throughout the water column. The seas, once turbid and stagnant, became clearer, and still richer in oxygen – enough to enable the evolution of larger life forms.[5]

The development of the anus had another consequence. Animals with a mouth at one end and an anus at the other have a distinct direction of travel – a 'head' in front and 'tail' behind. At first these animals lived by picking scraps off the thick carpet of slime that had lain on the ocean floor for more than 2 billion years.

Then they began to burrow beneath the slime. And then they ate the slime itself. The unchallenged reign of the stromatolites was over.

And when the animals had eaten all the slime, they started to eat one another.

There was still the small matter of worldwide glaciation to contend with. But evolutionary change thrives on adversity. Seaweeds flourished, providing more nourishing fare for early animals than bacteria.[6]

And it could have been that animal life was pushed in

the direction of increasing complexity by the very severity of the Snowball Earth glaciations. Following the maxim that what doesn't kill you makes you stronger, animal life had, at its dawn, to be resilient to survive the most demanding period of adversity in its history. Once the glaciations receded – as all glaciations in the history of the Earth eventually have – they left animal life leaner, meaner and ready to take whatever the Earth could throw at it.

Animal life burst into visibility sometime around 635 million years ago, in what is known as the Ediacaran period. This first flush of complex animal life was a flourish of beautiful, frondlike forms, many of which defy categorization.[7] Although some were animals, others may have been lichens, or fungi, or colonial creatures of uncertain affinity – or something so entirely foreign that we lack any means of comparison.

One of these, a strikingly beautiful creature called *Dickinsonia*, was broad but pancake-flat and segmented. It's easy to imagine one gliding gracefully over the sediment, much as flatworms or sea slugs do today.[8] Another fossil called *Kimberella* might have been a very early relative of molluscs.[9] Others, the rangeomorphs, are even harder to categorize. They resembled plaited loaves and probably remained in one place throughout their lives, although – like strawberry plants – they budded off new colonies round about the parent.[10] The world of these strangely beautiful, alien creatures was placid and quiet. They lived in shallow seas and dotted the shoreline amid the seaweed.[11]

The earlier Ediacaran creatures tended to be of this soft-bodied, frondlike sort. Creatures that looked more definitely animal-like, and which could have moved about, appeared somewhat later, after about 560 million years ago – alongside a widespread appearance of what are known as trace fossils. Trace fossils aren't impressions of creatures themselves, but signs of their activities. They include track marks and burrows. Trace fossils are as intriguing as the footprints of criminals that have just left the scene of the crime. We can say something of the criminal's build from a footprint, and even of their intent. But we can't say anything much about, say, the clothes they were wearing, or the weapons they were carrying. To do that you'd have to catch the criminal in the act. Rarely, very rarely, can we do the same for trace fossils. One such is a fossil called *Yilingia spiciformis*, which lived at the very end of the Ediacaran. Specimens are occasionally found at the end of their own trails, and they appear to look like the kind of segmented worms fishermen today use as bait.[12]

These traces are of incalculable importance. They are an echo, or an after-image, of a moment in evolution when animals first began to move around. Until that point, creatures were usually rooted to one spot, for at least some part of their life cycle. Tracks and traces are almost always left by animals accustomed to directed, muscular motion. If one's sources of food are all around, there is no need to go looking for it in one place rather than another. However, if an animal has a single direction of travel, with a mouth at one end, it is usually looking for something, and that something is food. At some time in the middle of the Ediacaran, animals actively started to eat one another. And once that was happening, they also started to find ways to avoid being eaten.

An animal burrowing in mud needs to have a dense, resilient body to enable it to penetrate the sediment. There are various ways of achieving this. The body of a burrowing animal could be braced by an internal skeleton, like, say, a Jack Russell terrier; or an external skeleton, like, say, a crab. External skeletons tend to start off soft and flexible (as in a shrimp) but may become hard and mineralized (as in a lobster). Another way is to organize one's body as a series of repeated segments, each full of fluid, and separated from the segments fore and aft by a kind of bulkhead. If the segments are contained in a tough external tube of muscle, you can essentially force yourself into the soil by exerting pressure on it. And if you move around like that, then you are an earthworm.

The marine relatives of earthworms do much the same, but many have flexible limb-like outgrowths on each segment that help them burrow, or row through the water, or crawl along the surface. Some of the earliest fossilized animal trails, such as those of *Yilingia spiciformis*, could have been made by worms like that.

Animals such as segmented worms have a more sophisticated organization than jellyfish or even very simple flatworms. And the crucial difference is that they have insides as well as outsides.

Jellyfish and simple flatworms, essentially, have no insides. Their guts are inpocketings of the surface, their connection to the outside serving as both mouth and anus.

More complex animals, by contrast, have a straight-through gut with a mouth at one end and an anus at the other. They may also have internal cavities that separate the through-gut from the external surface. It is in this space that internal organs can develop.

In general, jellyfish-grade animals lack such storage space. The presence of internal space means that the growth of the gut and the external surface are no longer tied together, allowing for the development of large, complex guts, and a larger size in general. Large guts and large size come in handy if your choice of occupation is the eating of your fellow creatures.

If such is your occupation, you will need teeth. And if you are to avoid being eaten, you will need armour. The animals of the Edenic Ediacaran were essentially soft, squishy and defenceless. The exile from Eden was harsh, merciless – and sparked by another of the Earth's great upheavals.

It happened during another period of heavy weathering, at the very end of the Ediacaran period. The Earth's crust took such a pounding from the weather that much of the land surface was eroded away, worn down to bedrock and dumped into the sea. This had two effects. First, the sea level rose markedly, drowning the coasts and making more space for marine life. The second was the sudden availability in the sea of chemical elements such as calcium, an essential ingredient for shells and skeletons.[13]

The earliest mineralized skeletons are about 550 million years old and belonged to an animal called *Cloudina*. These looked like stacks of very small ice-cream cones, nested within one another.[14] Fossils of *Cloudina* are found worldwide, and already at this early date, some of them show evidence of having been drilled into by some unknown but sharp-tongued predator.[15] Slightly later, at around 541 million years ago, a trace fossil called *Treptichnus* appears widely in the fossil record. *Treptichnus* is a specific kind of burrow in the sea floor, made by animals unknown. It marks the beginning of the Cambrian period, and the second great efflorescence of animal life – animals that burrowed, swam, fought and consumed one another. They had hard skeletons reinforced by calcium compounds. They also had teeth.

Perhaps the best-known animals from the Cambrian period are the trilobites. These were arthropods[16] – that is, animals with jointed limbs – that looked rather like pill bugs or woodlice. They were common in the seas from just after the start of the Cambrian to the Devonian, when they went into a decline. They eventually became extinct at the end of the Permian, some 252 million years ago.

Trilobites are relatively common as fossils. Every rock hound will have at least one in their collection, but their familiarity and ubiquity should not lead us to underestimate them. Trilobites were exquisitely beautiful, and as complex as any animal alive today. They had exoskeletons that they could moult as they grew, just like arthropods do today, from the tiniest midges to the largest lobsters. Perhaps most remarkable were their eyes, each one a collection of dozens, even hundreds of individual facets, like the eyes of a dragonfly. Each facet has been preserved in fossils as crystalline calcium carbonate. There was variation, of course. Some trilobites had enormous eyes, whereas others were blind. Some trilobites specialized in fossicking around on the sea floor, whereas others were better at swimming.

But there was more to Cambrian life than trilobites.

One day, around 508 million years ago, in what is now British Columbia, a mudslide swept part of the ocean floor down to greater depths – along with everything living in, on or above it. The animals were buried intact, in almost oxygen-free conditions. This rapid interment ensured that the animals remained whole. Even fine details of their soft tissues were left almost untouched by the subsequent half-billion years or so. During this time the rocks were compressed, very slowly, into shale, and, in the past 50 million years or so, thrust up from the oceans to rest among North America's highest peaks, where, since they were discovered in 1909, they have become known to us as the Burgess Shale. The creatures buried within represent a rare snapshot of ancient sea-floor life in the Cambrian period.

And what a menagerie it is. A cavalcade of spiny, jointed limbs, clattering claws and feathery feelers, all attached to animals obscurely related to today's crustaceans, insects and spiders. Some of these creatures were very strange, even when considered against the exuberant diversity of arthropods today. There was *Opabinia*, with its five stalked eyes and peculiar grasping jaws set on the end of a flexible hosepipe-like snout.

There was *Anomalocaris*, a metre-long predator, cruising the depths in search of prey it could stuff into its circular, garbage-grinder mouth with its sharp pincers.[17]

And most of all, *Hallucigenia*, a wormlike creature that

crawled on the sea floor, protected from above by the double row of long and unwieldy spines it bore on its back.

As arthropods crawled on the sea floor or swam above it, a wonderland of worms writhed in the ooze below.

Many of the creatures found in the Burgess Shale are only distantly related to animals alive today.[18] However, it's possible to discern to which of the many major animal groups each fossil is related, even if only as a remote and eccentric cousin. As well as arthropods – in their broadest sense, including *Hallucigenia* as well as fossils looking like the modern 'velvet worms' that stump around in the leaf litter of tropical-forest floors, each one looking like an earthworm but with stumpy, Michelin-Man legs – there were quite a few animals that were related to various kinds of worm that burrow in sediment.

As for the arthropods, so for the molluscs, which are as squishy as arthropods are spiky, at least on the inside. *Wiwaxia* combined the body of a segmented worm with the horny tongue, or radula, of a mollusc – the same radula which, in modern slugs, wreaks havoc on your lettuces. All clothed in a most unsluglike chain-mail suit.[19] Another animal with a radula, but which otherwise looked like an airbed crossed with a coffee grinder, was *Odontogriphus*. This, too, was a relative of the earliest molluscs.[20]

Elsewhere, there was *Nectocaris*, a very primitive, shell-less squidlike creature, and the earliest known member of the cephalopod molluscs.[21] Today, this group includes, in the octopus, one of the most intelligent and strangest of all invertebrates; and, in the colossal squid, the largest. The fossil history of cephalopods is just as majestic as their modern representatives would suggest, with the evolution

– not long after *Nectocaris* – of the nautiloids, squid with trumpet-like shells several metres long; and, eventually, in the age of the dinosaurs, the coiled ammonites, some of which grew as large as truck tyres, gracefully cruising the oceans.

Since the discovery of the Burgess Shale, similar deposits have been found, of broadly similar ages. They include the Chengjiang Fauna of southern China, and range across the globe from southern Australia to northern Greenland. All are remarkable in the fidelity of their fossil preservation, down to the finest details. The Chinese shrimplike fossil *Fuxianhuia*, for example, is known in such detail that it is possible to work out the nervous wiring in its brain.[22]

Such amazing preservation is extremely rare. It is the result of a perfect storm of geological circumstances and the biochemistry of interment. In almost all cases when fossils are found at all, they are only of hard parts already infused with minerals: shells, bones and teeth, rather than nerves, gills or guts. Fossils of the same approximate age as the Burgess Shale had been known for a very long time, but are all of the hard and shelly sort: a legacy of the sudden infusion of minerals into the sea at the end of the Ediacaran, which allowed animals to clothe themselves in armour.

The efflorescence of life forms that occurred in the Cambrian over the course of just 56 million years is unmatched by anything before, save the origin of life itself – nor, it has to be said, since. Although 56 million years is a long time, the subsequent 485 million years have only seen elaborations on what have become well-worked themes. It is, for example, less than the interval of 66 million years that has elapsed since the extinction of the dinosaurs.

It is not for nothing that this seismic upheaval in evolution has become known as the Cambrian 'Explosion'. However, it was less a sudden detonation than a slow rumble. It began with the break-up of Rodinia and the evolution and eclipse of the weirdly beautiful Ediacaran fauna, and continued until around 480 million years ago.[23]

By the end of the Cambrian period, all the major groups of animals still around today had made their first appearance in the fossil record.[24] Not just arthropods and various kinds of worms, but echinoderms (spiny-skinned animals such as sea urchins) and vertebrates (the backboned animals, which includes ourselves). One of the very earliest was the fishlike *Metaspriggina*, found in the Burgess Shale. Rather than having external calcite armour, it had an internal, flexible backbone, to which powerful muscles were anchored. All the better for swimming – and fast, to avoid the nightmarish pursuit of giant arthropods such as *Anomalocaris*.

Metaspriggina was one of the very first fish to have made an entry in the fossil record. And its story belongs to the next chapter.

Timeline 3. Complex Life

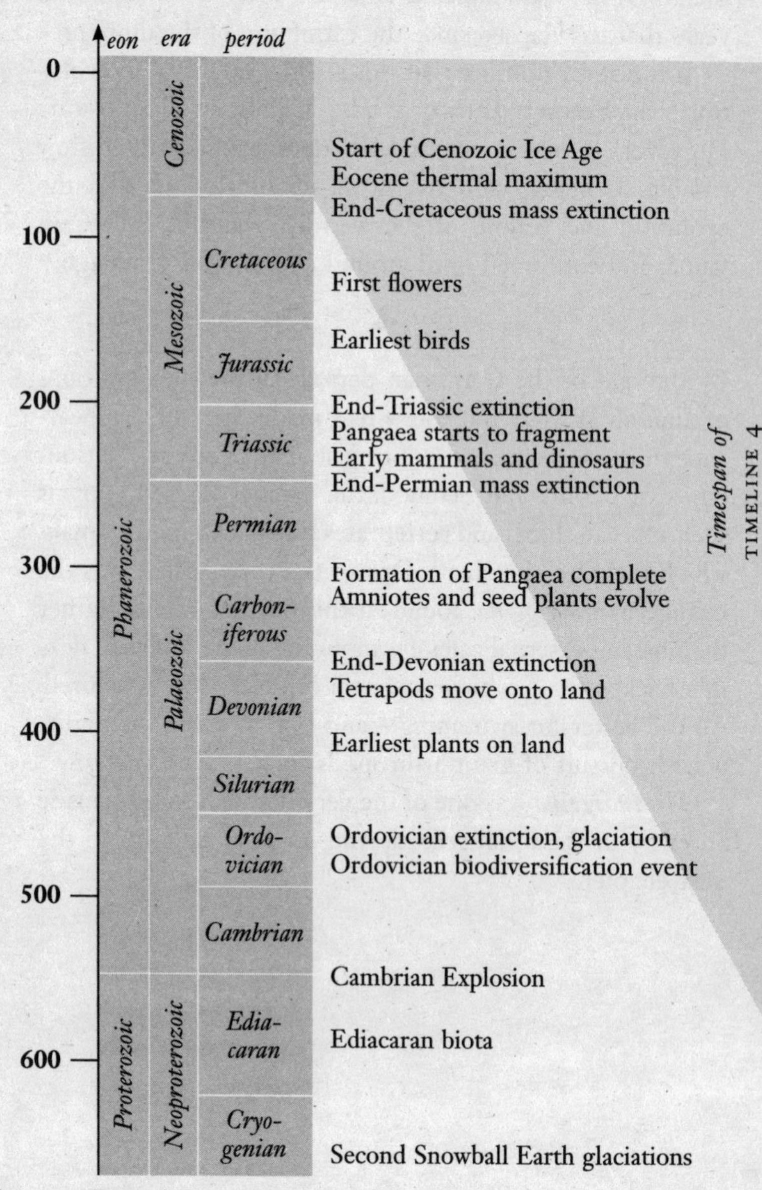

eon
era
period
0
100
200
300
400
500
600
Cenozoic
Mesozoic
Palaeozoic
Neoproterozoic
Phanerozoic
Proterozoic
Cretaceous
Jurassic
Triassic
Permian
Carbon-
iferous
Devonian
Silurian
Ordo-
vician
Cambrian
Edia-
caran
Cryo-
genian
Start of Cenozoic Ice Age
Eocene thermal maximum
End-Cretaceous mass extinction
First flowers
Earliest birds
End-Triassic extinction
Pangaea starts to fragment
Early mammals and dinosaurs
End-Permian mass extinction
Formation of Pangaea complete
Amniotes and seed plants evolve
End-Devonian extinction
Tetrapods move onto land
Earliest plants on land
Ordovician extinction, glaciation
Ordovician biodiversification event
Cambrian Explosion
Ediacaran biota
Second Snowball Earth glaciations
Timespan of
TIMELINE 4
(page 144)
Ages in millions of years
before the present

3

THE BACKBONE BEGINS

As the warm, shallow oceans of the early Cambrian were filled with the spiky clatter of arthropod pincers, events were afoot in the sandy slough of mineral grains below. A small creature called *Saccorhytus*, no bigger than a pinprick, was making a modest living filtering detritus from the water between the grains.[1] Filter-feeding was nothing new – sponges had been doing it for 300 million years – and many other creatures, such as clams, were reinventing it. Panning the sediment for edible morsels is a cheap and efficient means of making a living, especially for small animals with few metabolic demands. *Saccorhytus* was just such an animal.

Shaped like a potato, albeit very much smaller, *Saccorhytus* had a large, circular mouth at one end, ready to welcome in a current of water, drawn in, sponge-fashion, by ranks of waving cilia. Along each side was ranged a line of pores, like portholes on each side of a ship, through which the filtered water exited. Inside, nets of sticky mucus trapped particles of detritus from the water current. Most of *Saccorhytus*' insides were given over to this mouth-and-portholes arrangement, known as the pharynx. The mucus was rolled up into a rope and swallowed by an internal gut. This, with all the rest of the animal's viscera, was packed in a relatively small space at the back. The anus was internal, and faeces were swept out through the portholes, along with sperm or egg cells, expelled by the parent to take their chances in the outside world.

But *Saccorhytus* was otherwise helpless, as prey to the whims of its environment as the mineral grains between which it lived. Countless animals were doubtless swallowed by undiscriminating filter-feeders such as sponges or clams, even if they were beneath the notice of larger predators. Some of *Saccorhytus*' descendants evolved their way out by becoming larger, more mobile, armoured, fierce – or combinations of all four.

Being bigger means that an animal is less likely to be swallowed whole – although it might run the risk of being pecked at, piecemeal. To avoid this fate, some animals evolved suits of armour. Many other animals had already done this by reinforcing their outer layers with calcium carbonate, which they extracted from the mineral-rich seas. Calcium carbonate is one of the most common minerals – it is what makes up calcite, chalk, limestone and marble. The Cambrian seas were rich in calcium carbonate, and, when sculpted by living things, it becomes mother-of-pearl: the shells of clams and crustaceans, the microscopic spicules of sponges and the armature on which is built the fantastic shapes of coral reefs.

Some of the armoured heirs of *Saccorhytus* created their own distinctive suits of chain mail, each link sculpted from a single crystal of calcite. In doing so they became the echinoderms – the spiny-skinned ones – the ancestors of the starfishes and sea urchins of today. All modern echinoderms have a distinctive body shape based on the number five, entirely different from any other animal. In the Cambrian, however, their shapes were more varied. Although some were bilaterally symmetrical, a few were triradial (that is, with a symmetry based around the number three), and yet others were completely irregular. All started with the mouth-and-portholes pharynx of *Saccorhytus*, though this was replaced over time by other modes of feeding. No modern echinoderm feeds like this.

Against predation, then, the echinoderms opted for a strategy of armoured defence. Another solution, however, is escape – to swim away from one's attacker as quickly as possible. This solution was adopted by another branch of *Saccorhytus*' descendants. Some of these extruded a swishing tail from the back end of the pharynx, all the better for swimming away – fast – from any potential threat.

This started with a long, stiff-yet-flexible rod, which evolved from a developmental offshoot of the gut. One can think of this structure, called the notochord, as like one of those sausage-shaped balloons entertainers twist into amazing shapes at parties. Although very flexible, the notochord could spring back to its original long, narrow form when not under tension. This property made the notochord suitable for anchoring ranks of muscles on either side, which alternately contracted and relaxed. This threw the body of the animal into a series of S-shaped bends that propelled it through the water. The muscles were coordinated by regularly spaced offshoots of a nerve running along the upper surface – the spinal cord.

Cambrian animals called vetulicolians look a lot like this.[2] A vetulicolian, no more than a few centimetres long, has a *Saccorhytus*-like pharynx to which a segmented tail is attached. Although some vetulicolians swam through open water,[3] they spent most of their time buried in sand, with only their mouths showing, quietly drawing in sediment. If threatened, however, they could swish their tails and swim rapidly out of danger, settling into a new location, and using their tails to dig a new refuge in the sand. Cousins to the vetulicolians were the yunnanozoans, in which the tail and pharynx had started to grow together. As well as growing backwards, the tail also extended forwards, over the top of the pharynx, and, eventually, enclosed it, to give a more fishlike shape.[4] *Pikaia*, a strange creature from the Burgess Shale, was of this kind.[5] Another such animal was *Cathaymyrus* from the Chengjiang Biota of China.[6]

On first glance, *Cathaymyrus* looked like an anchovy fillet. Although its notochord and blocks of muscle were easy to see – the front part enclosing the pharynx – it lacked a great deal. A single pigment spot at the prow served for eyes. It was *sans* head, *sans* scales, *sans* ears, *sans* nose, *sans* brain – *sans* nearly everything. The Wizard of Oz should have found in it a willing customer: yet it had plainly refused any invitation to join Dorothy and her friends as they marched down the Yellow Brick Road. Nevertheless, *Cathaymyrus* and its relations have lived successfully if modestly for half a billion years, buried, tail first, in the unregarded interstices of the world, where they have spent almost all their time filtering detritus from seawater in time-honoured fashion. Only when threatened do they dare dart forth, swimming until they find safer refuge. A few of *Cathaymyrus*' relations have survived to the present day, and they are known as lancelets, or the amphioxus.

Cathaymyrus combined pharynx and tail into a single, streamlined animal. Some of its cousins, though, adopted an entirely different mode of life. Rather than integrate pharynx and tail, these creatures – the tunicates – deconstructed them, using each to its best advantage at a different phase of life.[7] The tunicate larva is mostly a tail, with a simple brain, eye-spot and gravity-sensing organ. These senses are rudimentary but serve its needs, which is to tell light from dark, and detect which way is 'down'. The larva has only a rudimentary pharynx and cannot feed. Such is entirely consistent with its purpose, which is to seek out a site, deep and dim, where it can settle as an adult. Once a suitable site is found, the animal sticks head first to the spot. Its tail is resorbed and the creature balloons into what is essentially a gigantic pharynx, devoted to feeding. Being stuck to one spot makes it easy prey, so tunicates have evolved their own suits of armour, in the form of a 'tunic' (whence 'tunicates') made of cellulose. This substance, otherwise found only in plants, is entirely indigestible. The tunics of tunicates may contain other exotic substances extracted from seawater, such as nickel, or vanadium, and are sometimes stiffened by minerals as well. The tunicate *Pyura*, for example, looks entirely like a rock, until one breaks it open. Tunicates have lived like this since the Cambrian.[8]

Tunicates have always fed by the age-hallowed mouth-and-portholes pharynx filtration system pioneered by *Saccorhytus*.[9] Their closest cousins, the vertebrates, went down an altogether different route. They transformed what once offered a way of escape – the notochord and tail – into a dedicated means of forward motion. *Cathaymyrus* and its relatives used its notochord-supported tail for very short bursts only. In tunicates, the notochord-stiffened tail evolved only in the larva, for the most part, where it was used, quite specifically, to find a good spot for settling down, and, once settled, there it stayed. Such animals needed only minimal information about where they were going. For them, the purpose of the tail was rapid departure on a journey soon ended.

No vertebrate, however, has ever spent any appreciable part of its life cycle fixed in one place.[10] Being constantly on the prowl demanded a much more comprehensive battery of senses. Vertebrates evolved large, paired eyes; a refined sense of smell; and an elaborate system for detecting water currents.[11] Vertebrates became much more sensitive to their environment, and their place in it, than any other members of the school of *Saccorhytus* – tunicates, the amphioxus, vetulicolians, echinoderms and so on. An elaborate sensory system required a complex, centralized brain. The brains of vertebrates matched or even exceeded the complexity of the brains of other highly mobile animals such as crustaceans, insects or that doyen of movement, the octopus, even if these brains were constructed along totally different lines.

And so it was that from the murk of the Cambrian sea floor emerged, like flickers of sunshine flitting through the water, fishes such as *Metaspriggina*,[12] *Myllokunmingia* and *Haikouichthys*.[13] These slivers of creatures are evidence that vertebrates had evolved and were widespread by the middle of the Cambrian. These, the earliest fishes, had mouths, but no jaws: and a pharynx, though it was no longer used for filter-feeding. Being much more active animals than their tunicate cousins, vertebrates needed a better supply of oxygen. The ancient pharyngeal portholes that traced their ancestry back to *Saccorhytus* were transformed into gill slits. Water entering through the mouth was squirted past the gills by muscular action. The feathery gills, rich in blood vessels, extracted oxygen from the water, and expelled carbon dioxide. Vertebrates, then, took the pharynx and hot-rodded it. Fields of gently beating cilia were replaced by ranks of muscles for ventilation, that is, breathing; and actively catching prey.[14]

Vertebrates require more energy than other animals in part because they are, in general, rather large. Whales and dinosaurs – vertebrates all – are the largest animals of any kind that have ever been, but they are not alone. Think of fishes such as whale sharks and basking sharks; reptiles such as the python and other boa constrictors, and the Komodo dragon; mammals such as elephants and rhinos. Few invertebrates can match them in size. We humans, too, are uncommonly large for animals.[15] It is true that some vertebrates can be very small, weighing a matter of a few grams: but *all* vertebrates are visible to the naked human eye. Many invertebrates, on the other hand, are hardly visible without a hand lens or microscope.[16]

Insects are the most numerous invertebrates, and support themselves with an external skeleton made of a flexible protein called chitin. When the insect needs to grow, it sheds its entire external skeleton, inflates itself, and waits for the newly made one, still rather soft, to harden, before it can move. This is one reason insects are small. More than a certain size, and an insect without an exoskeleton would be crushed under its own unsupported weight. The close cousins of the insects, the crustaceans, also moult, but they live mainly in water, which supports their weight. This means that crustaceans can grow somewhat larger than insects. Think, for example, of crabs or lobsters, which can grow much larger than any insect. Nonetheless, even the largest lobster is a tiddler compared to many vertebrates.

The most primitive vertebrates alive today are lampreys and hagfish. These have no external armour, and have probably been that way since they first evolved. Like *Metaspriggina* and other extremely early fishes, they also lack jaws and paired fins. Other vertebrates, though, adopted thick coats of armour plating. Armoured fishes appeared later in the Cambrian. Although they still lacked jaws, and were supported internally by a notochord, most early fishes were clothed in a suit of armour.[17] This was often a set of solid plates around the head and pharynx, but loose and scalier at the back end, to allow the tail to move. The armour was not made of calcite, or calcium carbonate, but a different mineral, hydroxyapatite, a form of calcium phosphate.

Calcium phosphate armour is unique to vertebrates in the animal kingdom.[18]

The armour of the earliest fishes was usually a variant of a thick layer-cake of hydroxyapatite done three ways. At the base was a spongy layer. In the middle was a somewhat denser variety. On the top was a thin layer of a very hard, very dense form of hydroxyapatite. These three forms are now known as 'bone', 'dentine' and, lastly, 'enamel' – the hardest substance made by any living organism. Nowadays, bone, dentine and enamel occur, in just these layers, in our teeth. When hard tissues first evolved in vertebrates, there were, in essence, teeth all over the body. Even today, the individual scales of sharks each take the form of a tiny tooth, which is why sharkskin is abrasive and was once used as sandpaper.

Vertebrates evolved armour for the same reason that other Cambrian creatures became clothed in hard tissues – as a means of defence.[19] The evolution of armoured fishes coincided with the appearance of predatory nautiloids, and gigantic ocean-going scorpions called eurypterids.[20] Perhaps the most terrifying eurypterid was *Jaekelopterus*, which lived in the Devonian. A nightmare with great big goggly eyes and huge pincers, it grew to about 2.5 metres and probably dined on fishes.[21]

The earliest group of fishes to become clothed in armour were the pteraspids. Although the head shields of pteraspids were sometimes extended on either side to act as hydroplanes, they had no flexible, paired fins. Thickly armoured on the outside, very little is known about what pteraspids looked like inside, as their braincases were made of cartilage, which decays easily, and they were supported internally by a spongy-yet-springy cartilaginous notochord. In some armoured fishes, however, the soft cartilage inside the head became mineralized, and this means that the shapes of the brain and the associated blood vessels and nerves have been preserved in great detail. These fossils show that these jawless armoured fishes were constructed along the same lines as lampreys – lampreys in armour.

Jawless armoured fishes thronged the seas from the latest Cambrian to the end of the Devonian, and came in a wide variety of extraordinary shapes. Some were boxed in plate armour and spent much of their time cruising along the sea floor or grubbing in the mud for detritus. In others, such as the stylish thelodonts,[22] the armour was a sharkskin-shagreen of more flexible chain mail, allowing for more rapid movement in open water.

The earliest fishes, such as *Metaspriggina*, had paired eyes close together right at the front, like motorcycle head-lamps. There was no space for a nose, or nostrils. Smell was the business of cells in the pharynx, a holdover from the ancient filter-feeding heritage of vertebrates. In pteraspid fishes, however, the eyes became displaced towards the sides of the head to make way for a nostril, which was single, and at the top of the head. The brain had become divided into left and right hemispheres, widening the face.[23]

The single nostril of pteraspids (as in lampreys) led to a single sense organ – the nasal sac – in contact with the base of the brain. Other jawless fishes, though, were evolving in a new direction. Fossils of the brain of one jawless fish, *Shuyu*,[24] show that it had two nasal sacs, opening into the mouth cavity, rather than a single nostril opening separately on the top of the head. This arrangement, widening the face still further, is entirely characteristic of jawed vertebrates, but of neither lampreys nor pteraspids. Some other advanced jawless fishes also sported paired pectoral fins (the pair just behind the head), something that neither lampreys nor pteraspids boast, but a typical feature of jawed vertebrates. The stage, then, was set for the evolution of jaws.

When armoured fishes evolved to cross that line they became an entirely new kind of animal.[25] Today, jawed species comprise more than 99 per cent of all vertebrates. Of the jawless vertebrates, only lampreys and hagfishes remain.

Jaws evolved when the first gill arch – the cartilaginous division between the mouth and the first gill slit – became hinged in the middle and scissored in half, backwards, to become upper and lower jaws. This resulted in the first gill slit being squeezed until it became a tiny hole, the spiracle, just behind and above the upper jaw.

The first vertebrates with jaws were the placoderms, and at first sight they looked much like any other armoured fishes, with thick, bony head shields. Closer inspection reveals, apart from jaws, other refinements only seen in jawed vertebrates, such as a second set of paired fins in addition to the pectorals. These were the pelvic fins, situated more or less on either side of the anus.[26] The placoderms originated deep in the Silurian period and thrived until the end of the Devonian.

The more primitive placoderms, the antiarchs, were as thickly armoured as any pteraspid. The more sophisticated placoderms, the arthrodires, in contrast, generally (but not always) carried lighter armour, and one of them – *Dunkleosteus*, which grew up to 6 metres long and had vast razor-sharp jaws, became the top predator of the Devonian ocean.

Note that I refer to the jaws of *Dunkleosteus*, not the teeth, because placoderms did not have any teeth that we would recognize.[27] The cutting surfaces in the formidable jaws of this creature were formed from the honed edges of the bones themselves.

One of the most advanced placoderms was *Entelognathus*, even though, at an age of 419 million years, deep in the Silurian, it is among the earliest known.[28] *Entelognathus* has the heavily armoured head and trunk characteristic of an arthrodire, though, at about 20 centimetres long, very much smaller than its monstrous cousin *Dunkleosteus*.

Another difference from *Dunkleosteus* – and every other placoderm – was that the jaws were fringed with bones recognizable by comparison with a modern bony fish: there was a distinct upper jaw (maxilla) and lower jaw (mandible). This creature, *Entelognathus*, was the earliest vertebrate that could crack a smile that we would recognize.

Although the placoderms did not survive the end of the Devonian, three other groups of jawed vertebrates arose from placoderm ancestors. They were the cartilaginous fishes (sharks, rays and their kin); the bony fishes (which include most modern fishes, from sturgeons and lungfishes to sardines and seahorses; and all land vertebrates, including ourselves); and another wholly extinct group called the acanthodians, or spiny sharks.

The acanthodians made it to the Permian before dying out. In most cartilaginous and bony fish, the notochord – the firm-yet-flexible strut that supports the body – is replaced, during development, by a segmented structure, the backbone. In cartilaginous fish, the backbone is, of course, cartilaginous, although it is sometimes mineralized to some extent. In bony fishes the cartilage is generally replaced by bone. It is not known whether placoderms or acanthodians had a backbone, rather than a notochord, though, if they had, it would have been cartilaginous.[29]

Acanthodians were scaly rather than armoured, and distinguished by having a prominent spine at the leading edge of each fin. Their internal anatomy, however, was entirely cartilaginous, and rather similar to that of sharks.[30] Acanthodians were an early offshoot of cartilaginous fishes, a group which survives and thrives to this day.

Living alongside *Entelognathus* in the Silurian seas was a fish called *Guiyu*. This was the earliest well-known member of the bony fishes, the group that includes the vast majority of all vertebrates today.[31] There were bony fishes living earlier than *Guiyu*, but their fossils are rather fragmentary and debatable. *Guiyu* is special, though, not because it is well preserved, or because it is a bony fish. It is special because it was among the very earliest fishes that belonged to a group known as the lobe-finned bony fishes, a peculiar offshoot of bony fishes that evolved into land vertebrates – and us.

4

RUNNING AGROUND

By now the oceans swarmed with creatures, from the exuberant burst of life in the early Cambrian to the fish-seething seas of the Devonian. But few organisms had yet dared to venture above the surface of the waters, onto dry land. And with good reason.

First, for a long time there was very little land. To begin with, the continents accreted slowly. When tectonic plates collided, arcs of volcanic islands arose. Thunderheads of magma from deep inside the Earth punched through the crust, making more. These islands were joined by others, and, shoved together by the restless planet beneath, became the first continents.

Second, life on land is hard. Water is a nurturing cradle. Without its buoyancy, creatures feel every gram of their own weight, dragging them down. Under a scorching sun, their tissues may soon dry out. Without a constant film of water, gills cannot function, so an animal is unable to breathe. Any brave venturer onto land would have been crushed, desiccated and asphyxiated. Pioneers on land would have found there an environment almost as hostile as empty space.

Pitiless it would have been, too, with no surface other than barren, volcanic rock. There were no trees to offer shade, because trees had yet to evolve. There were no soils apart from dust scoured by the wind, because it is the action of living things – roots, fungi, burrowing worms – that creates and enriches the soils in which plants can grow. Earth above the waterline was a desert as dry and lifeless as the surface of the Moon that still loomed large on the horizon.

But life, as we have seen, has a tendency to rise to challenges. An entirely new environment, free from competition in the bustling ocean, offered untapped opportunities for diversity and growth to those creatures that could find ways to tame it. The first step was the colonization of inland ponds and streams by algae, which happened at least 1200 million years ago.[1] Perhaps even then, crusts of bacteria, algae and fungi hid in sheltered nooks along the barren seashore. It is possible that some of the more frond-like Ediacaran animals spent some time above the waterline, if caught between the tides.[2] In the Cambrian, a creature unknown slithered onto the low, sandy beaches of the continent of Laurentia,[3] leaving trails that look uncannily like motorcycle tyre tracks.[4] But these were actions of bravura defiance, as if the motorcyclist had performed a few wheelies before seeking sanctuary once more beneath the waves. Life had ventured onto land, but had not yet come to stay.

The invasion of the land started in earnest in the middle of the Ordovician period, around 470 million years ago[5] – about the same time as a spurt of evolutionary innovation in the sea in which many of the strange creatures of the

Cambrian were replaced by those of a more modern cast.[6] Small, creeping plants such as liverworts and mosses made millions of miniscule bridgeheads onto land. It was their spores, tough and resistant to desiccation, that allowed them to be any more than occasional visitors. The first trees reached for the skies soon afterwards. The first were the nematophytes. One, *Prototaxites*, had a trunk more than a metre in diameter, and grew several metres tall. It was less a tree, or even a tree fern, than a giant lichen – a fungus associated with an alga.

Beneath everything, the Earth still moved. An episode of volcanic eruptions spewed out rocks that reacted well with carbon dioxide, scrubbing it from the atmosphere. Without carbon dioxide to stoke the greenhouse effect, the Earth cooled. At the same time, the giant southern continent Gondwana moved over the South Pole. Glaciers once again formed on land. The glaciers sucked water from the sea, reducing sea levels. This shrank the space on the continental shelves where most animals lived. This ice age lasted about 20 million years, from 460 to 440 million years ago. It was not as cataclysmic as the one that saw in the Ediacaran, still less the one that fuelled the Great Oxidation Event. However, many species of marine animals died out.

Life, as ever, responded to the changing environment. After the glaciation, hardy, fernlike plants appeared, with even more desiccation-resistant spores than liverworts. Liverworts, outcompeted, were driven to the damp, shady places where they still live today. The land, once naked, became clothed in a brilliant green.

By the late Silurian, around 410 million years ago, there were woodlands of nematophytes, mosses and ferns. The roots of the plants began to grind the rocks beneath them, to make soil. With soil evolved the soil fungi, and some of these – the mycorrhizae – linked up with plants to form beneficial associations. The fungi spread out in the soil, mining it for minerals important for the growth of plants. In return, the plants offered food, made by photosynthesis. Plants with mycorrhizal extensions on their roots would prove to do much better than plants without. Today, virtually every plant grows thanks to a mycorrhiza lurking in the soil around its roots.[7]

Exposed to wind and weather, the plants shed scales, and spores, and other matter, and, in the damp spaces in the forest litter, small animals began to crawl.

The first animals on land were small arthropods – centipedes; spiderlike animals such as harvestmen; and springtails, close cousins to the insects that would soon evolve and become the most successful land animals the Earth has ever seen, both in terms of numbers of individuals, and of species.

Throughout the Devonian, the forests grew and spread. The forests would not have looked much like the forests of today.[8] Early forest trees, the cladoxylopsids, for example, were more like giant reeds, shooting hollow, branchless stems 10 metres or so into the sky, terminating in brush-like structures like fly whisks.[9] Later additions included plants similar to clubmosses and the field horsetail, *Equisetum*, found in damp places even today. The modern forms are very small; their ancient relatives were giants.

The clubmoss *Lepidodendron* grew up to 50 metres tall; the horsetails, 20 metres. Most of these trees were hollow. They contained no heartwood, and were supported by their thick external rinds. Some of the trees, such as *Archaeopteris*, looked more like modern trees, and had heartwood – except that they shed spores as ferns do, rather than reproducing with seeds.

This wealth of plants would seem, at first, a source of food too good to miss. But for millions of years, plants were off the menu for animals. Woody tissue is tough and indigestible, and the plants also produced chemicals such as phenols and resins that animals were unable to tolerate. Plant material could only be eaten once broken down into digestible detritus by bacteria and fungi. For a very long time, plants were not so much a food source as the backdrop for miniature dramas, as tiny carnivores hunted miniature detritivores beneath the leaf litter. Herbivory was a skill that had yet to be developed. First, by insects that started to feed on the sensitive parts of plants – reproductive structures such as cones. And then, by an entirely new arrival from the sea – the tetrapods.

Animals, like all life, originally evolved in the sea. Most of their descendants are still there, and vertebrates are no exception. Most vertebrates, even today, are fishes. With this perspective, the tetrapods – those vertebrates that have made the move onto land – can be seen as a rather strange group of fishes that have become adapted for living in water of negative depth.

Their roots go back to the Ordovician, when the first

jawed fishes emerged, as part of the great increase in biodiversity at that time.[10] By the Silurian, many jawed fishes had appeared, such as *Guiyu*, which we met in chapter 3. In these early fishes are combined features nowadays seen in two quite separate groups. The first, the ray-finned bony fishes, includes virtually all fishes alive today, from groupers to gouramis, trout to turbot. In these fishes, the paired fins are anchored directly to bones in the body wall. They were not always so dominant. In ancient times, it was their cousins, the lobe-finned bony fishes, that ruled. As the name suggests, the paired fins in these fishes are held clear of the body by stout fleshy extensions, supported by extra bones.

The lobefins were once a varied group, including among their number the onychodonts, creatures with loose-boned skulls and peculiar, tusklike teeth; and the gigantic, predatory rhizodonts. The largest rhizodont, *Rhizodus hibberti*, grew up to 7 metres in length. In between was a variety of forms, many of which were covered in thick scales coated with a version of enamel.

Perhaps the most conservative lobefins were (and still are) the coelacanths. These appeared in the Devonian[11] and looked more or less the same until they disappeared during the age of the dinosaurs – or so for long it seemed. In 1938, a specimen, recently deceased, was discovered off the coast of South Africa, one of a population that still lives near the Comoro Islands, in the Indian Ocean.[12] More recently another population was found in Indonesia.[13] These animals seem hardly changed from their remote, Devonian forebears. Although known to artisanal fishermen, they may have eluded scientific notice because of their habitat, which is in deep water close to vertical, submarine cliffs.

Some of the lungfishes, in contrast, have evolved almost beyond recognition. Although the Australian lungfish, *Neoceratodus*, is a scale-armoured freshwater fish that looks much like ancient lobefins, its cousins, the South American *Lepidosiren* and the African *Protopterus*, have changed so much that they have in the past been confused with tetrapods.[14]

The clue is in the name.

Although all fish started out with lungs – originally a pouch that grew out of the roof of the mouth – in most fish it has become separate, a gas bladder used to regulate buoyancy. In the coelacanth, which is exclusively marine, it is filled with fat. Lungfishes, however, live in rivers and ponds that can dry out, leaving the fish literally out of water. As a consequence, lungfishes make much more use of the lungs to breathe air directly. Indeed, *Lepidosiren* must breathe air to survive. This does not mean that lungfishes are especially closely related to tetrapods. Their adaptations to land have been made independently, and, in *Lepidosiren* and *Protopterus*, the limbs have withered away to thin whip-like structures, rather than becoming stout enough to support the animal's weight on land. The earliest lungfishes, from the Devonian, were much like other lobefins of their age.

So, too, were the fishes whose cousins eventually made the move onto land. Creatures such as *Eusthenopteron* and *Osteolepis* were as fishy as you please, but their close cousins were already evolving to a state where life above water would be an occasional indulgence, and then a regular habit.

Many of these fishes lived in shallow, weed-choked waterways, where they preyed on their smaller relatives. Some became large, and used their flexible, bone-supported fins to pick their way to the best spots where they could ambush unsuspecting passers-by. Many rhizodonts were like this. Another group, the elpistostegalians, went considerably further.

Elpistostegalians were every inch the shallow-water predator. Unlike most fishes, which tend to be compressed from side to side, they were flattened up and down, like crocodiles – all the better for lurking in shallow water. Some even had eyes placed on top of the head, rather than at the sides, to complete the picture. Their unpaired fins – dorsal, anal and so on – were reduced or absent altogether, and their paired fins developed into what, for all practical purposes, were small arms and legs, with fin-like fringes. The late Devonian *Tiktaalik*[15] was a typical example; *Elpistostege*[16] was another. These animals were a metre or so in length, about the size and shape of small crocodiles. They had broad, flat heads with the eyes on top and in the middle, a sinuous body and stout, leg-like forelimbs. The bones in the limbs correspond in detail with those in land vertebrates. These fishes had lungs, and probably didn't use their internal gills very much. The part of the skull roof that would normally have extended over the gill region is rather short and formed a distinct 'neck', all the better for an ambush predator that needed to turn its head rapidly to grasp fast-moving prey.

Elpistostegalians were tetrapods in almost every respect, except for the fringe of fins that adorned their legs, in place of digits – fingers and toes.

Tiktaalik, *Elpistostege* and their cousins lived around 370 million years ago, near the end of the Devonian. Their history, though, went much further back. One of their number had traded fin rays for digits at least 25 million years earlier. Some 395 million years ago, one of them left its footprints on a beach in what is now central Poland.[17] Nobody knows what kind of tetrapod created those tracks, but nothing but a tetrapod could have made them.

Apart from their early date, what's striking is that they were not made in fresh water, but on a tidal flat, near the sea. The very earliest tetrapods, like Venus,[18] emerged directly out of the ocean. They were adapted to salt water, or perhaps the more brackish water of estuaries.[19]

And still, beneath it all, the Earth moved. Ever since the break-up of the supercontinent Rodinia, the continents had been scattered, separate. Slowly, the half-billion-year tide of continental drift was starting to turn. The Ordovician extinction, when the great southern continent of Gondwana moved over the South Pole, was a harbinger of things to come.

Towards the end of the Devonian, Gondwana and two great northern landmasses, Euramerica and Laurussia, had started to grind their way towards one another. The collision would produce enormous ranges of mountains, and a single, vast landmass – Pangaea. The coalescence of the continents, once again, made its effects felt on the creatures that lived on its surface: much in the way that bedclothes, when shaken, will displace the toys and crumbs, books and breakfast things carelessly placed upon them. The action of the weather on the raw new mountains sucked carbon dioxide from the air, reducing the greenhouse effect, and prompting a return of glaciers over south polar Gondwana. Elsewhere, volcanism took its toll. Once again, extinction beckoned.

Most of the extinctions happened in the sea. Corals were badly hit. Reef-forming sponges called stromatoporoids, common in the Devonian, were driven to extinction.[20] Stromatolites made a resurgence on the reefs. The tumult spelled doom for the last of the armoured, jawless fishes; for the placoderms; and most of the lobefins, too. But other groups survived. The closing epochs of the Devonian were marked by a diversity of tetrapods.

At first though, tetrapods stayed very much in the water. Even though they had limbs, with digits, they occupied aquatic ambush-predator niches similar to those of the rhizodonts and elpistostegalians, which they replaced. Whatever limbs with digits were for, they had not evolved specifically for life on land.

Among the most primitive tetrapods were *Elginerpeton*[21] from Scotland, and *Ventastega*[22] from Latvia. There was *Tulerpeton*[23] and *Parmastega*[24] from Russia, and *Ichthyostega*, from the tropical swamps of what is now eastern Greenland. *Parmastega* looked and lived much like *Tiktaalik*, or a modern caiman, cruising the water with only its eyes visible above the surface. *Ichthyostega* was biggish – about a metre and a half in length – heavily built, and with a curiously shaped backbone suggesting that if it moved on land, it flopped around like a seal, rather than using its thick and stumpy legs.[25] *Acanthostega*, also from Greenland, was half the length of *Ichthyostega* and much slenderer in build. Although it had limbs, these stuck out at the sides and were entirely the wrong shape for walking anywhere. It had internal gills – just like a fish – so it would have been totally confined to the water.[26] In contrast, its contemporary, *Hynerpeton*, from Pennsylvania, was well-muscled and quite capable of life on land.[27] By the end of the Devonian, tetrapods had become a very diverse but primarily aquatic group of strange lobe-finned fishes, with legs.

One might get the impression, however, that the earliest tetrapods were not very serious about legs, or, at least, hands and feet. *Tulerpeton* had six digits per limb; *Ichthyostega*, seven; *Acanthostega*, no fewer than eight.[28] Many tetrapods have since lost digits through evolution, and even entire limbs, but no tetrapod today develops normally with more than five digits per limb. The five-digit limb (a state known as pentadactyly) seems so ingrained that one might imagine it as an archetype in the mind of God, the occasional six-fingered creature an offence against the natural order.

The earliest flush of tetrapod diversity survived the end of the Devonian, but was gradually replaced during subsequent

Carboniferous times by a more 'modern' fauna of smaller, slenderer creatures.[29] These looked more salamander-like than fishy, and had settled on how many digits they should sport at the end of each limb.

Around 335 million years ago, when Pangaea was becoming welded into its final form, the dark, steamy forests of what is now West Lothian, Scotland, were alive with creepy-crawlies and the croaks of early tetrapods, in a volcanic environment and perhaps associated with hot springs. So much so that one of the tetrapods from that rich seam was named *Eucritta melanolimnetes* – the Creature from the Black Lagoon.[30]

♣

Even when they had evolved legs sufficiently strong to bear their weight on land, there was one aspect of the lives of the earliest tetrapods that tied them to the water – reproduction. Like modern amphibians, these early tetrapods relied on returning to the water to breed. Their young would have been like tadpoles – finned, fish-like creatures, with gills for breathing.

However, a group of animals was about to emerge that would revolutionize reproduction and enable the final conquest of the land. Living in the coal forests, amid the croaks of other early land vertebrates, the scuttle of scorpions the size of large dogs, and the menacing presence of giant eurypterid sea scorpions that had followed the tetrapods ashore, was an animal called *Westlothiana*. This small, lizard-like creature[31] was evolutionarily close to the ancestry of a group of tetrapods that evolved eggs with hard, waterproof shells. Each egg a private pond, they could be laid away from water, finally severing the connection between vertebrate life and the sea.

These were the animals which would, one day, evolve into reptiles, birds and mammals.

5

ARISE, AMNIOTES

The *Archaeopteris* and cladoxylopsid forests were swept away by the extinctions consequent on the formation of Pangaea. The corals and sponges that had built the great reefs of the Devonian oceans were wiped out. All the armoured fishes – the placoderms – died out, along with most of the lobe-finned fishes and all but a few trilobites. The scum, the slime, the angel-hair threads of cyanobacteria, moved in and took over. As they had in the earliest times, stromatolites ruled the reefs, at least for a while.[1]

The extinctions were a setback for the earliest tetrapods, whose first brave forays onto land were, quite literally, stopped in their tracks. Those tetrapods that survived the extinction stayed close to water, and, preferably, in it.

There were some, however, that regrouped, and attempted to reconquer the land under the rawness of the sky. These were a very different breed from the earliest tetrapods, creatures that were, in the great scheme of things, hardly more than fish with legs.

At the very start of the Carboniferous, a metre-long, superficially salamander-like creature called *Pederpes* crawled ashore.[2] Unlike the polydactylous extravaganza of the earliest tetrapods such as *Acanthostega* and *Ichthyostega*, *Pederpes* had established the pattern that would last to this day, of no more than five digits per limb – although its fossil remains hint that it maintained a vestigial sixth digit, a memorial of times past.

Pederpes was a relative giant for its age, though. It shared its world with many much smaller tetrapods[3] that prowled the water margins for small arthropods such as millipedes, or fought tiny to-the-death battles with scorpions – and somewhat larger-scale engagements with such eurypterids as came ashore, following in the footsteps of their age-old prey.[4] These earliest Carboniferous tetrapods, though much better adapted for land life than their Devonian kin, did not stray far from water. They lived on floodplains that were often inundated. The journey onto land had taken a few steps forward, but was still tentative, provisional.

Some early Carboniferous tetrapods, however, remained aquatic. A few contrived to lose the limbs they had so lately acquired. *Crassigyrinus* – a metre-long moray-eel-like predator with tiny limbs and huge jaws stuffed full of teeth, was an aquatic menace that stalked early Carboniferous rivers and ponds. A few went further. Small serpentine amphibians called aïstopods lost their limbs entirely.[5] These creatures were throwbacks to a vanished age, tetrapods that had never left the water at all. The tetrapod commitment to land was, for many millions of years, no more than equivocal.

The land plants that shaded the tetrapods in the wake of the end-Devonian extinctions were, like the tetrapods themselves, small and weedy compared with their ancestors. It took time for the forests to recover, but when they did, they became the mightiest rainforests the world had ever seen. They were dominated by 20-metre-tall horsetails such as *Calamites*, and lycopod clubmosses such as *Lepidodendron* that soared 50 metres into a sky that was not blue but brown, and filled with the stench of burning.

Most trees nowadays grow slowly, and live for decades, even centuries. Their bodies are supported by a core of wood. Closer to the bark, columns of vessels transport water upwards to the leaves, to fuel photosynthesis; and freshly minted sugars downwards, to feed the roots and the rest of the plant. Each tree will reproduce many times in its long life. In the rainforest, the leaves in the canopy shadow much of the ground beneath, and create another, entirely separate ecosystem high above the dim forest floor, of plants and animals that rarely, if ever, touch the ground.

The Carboniferous lycopod forests were not like this at all. The lycopods, like their Devonian forebears, were hollow; supported by thick skin rather than heartwood, and covered in green, leaflike scales. Indeed, the entire plant – the trunk and the crown of drooping branches alike – was scaly. With no columns of vessels to transport food, each one of the scales was photosynthetic, supplying food to the tissues close by.

Even stranger to our eyes, these trees spent most of their lives as inconspicuous stumps in the ground. Only when it was ready to reproduce did a tree grow, a pole shooting upwards like a firework in slow motion,[6] to explode in a crown of branches that would broadcast spores into the wind.

Once the spores had been shed, the tree would die.

Over many years of wind and weather, fungi and bacteria would etch away at the husk until it collapsed onto the sodden forest floor below. A lycopod forest looked like the desolate landscape of the First World War Western Front: a craterscape of hollow stumps, filled with a refuse of water and death; the trees, like poles, denuded of all leaves or branches, rising from a mire of decay. There was very little shade, and no understorey, apart from the deepening litter forming around the shattered wrecks of the lycopod trunks.

The profligate lives of lycopods had immense consequences for the entire world. The rapid and repeated growth of lycopod trees used an incredible amount of carbon, all of it derived from carbon dioxide in the atmosphere. This extravagant consumption – along with the intense weathering of the new-made mountains – contributed to the lessening of the greenhouse effect and the renewed growth of the glaciers around the South Pole.

Secondly, most of the creatures that are responsible for the dismemberment of dead trees today – termites, beetles, ants, and so on – had yet to evolve. There were as yet few animals capable of eating plant matter. Among the select few were the palaeodictyoptera, one the very first groups of insects to evolve wings and fly. Some of these animals were as large as crows, and had not two pairs of wings, as modern flying insects do,[7] but three. In front of the usual two pairs was a pair of small, vestigial flaps, remnants of a still earlier age of many-winged insect flight that is now lost. They also had prominent, sucking mouthparts, like bugs. Flying high above the ground, they alighted high on the lycopods to dine on their tender, spore-producing organs.[8]

Third, all that photosynthesis produced enormous quantities of free oxygen. There was, in fact, so much oxygen in the atmosphere that lightning strikes could set trees ablaze like torches, even in a waterlogged swamp forest, producing masses of charcoal, and rendering skies permanently brown and smoky.

The charcoal, the rapid burial, the imperceptible rate of rot, meant that many lycopod trunks were swiftly entombed entire in the forest floor, and, 300 million years later, emerged as coal. This is the coal that gives its name to the entire age – the Carboniferous – even though coal forests persisted well into the Permian. Around 90 per cent of all known coal deposits were laid down in just one heady 70-million-year interval, the age of the lycopod forests.[9]

It was a world in which amphibians thrived, evolving into a variety of forms. As the smaller ones writhed and burrowed into the banks, chasing small scorpions, spiders and harvestmen, their larger cousins remained aquatic, cruising the water for smaller prey, or snapping at the giant mayflies, the palaeodictyopterans, the gull-sized dragonflies and other winged insects that chanced to alight on the water surface.

Some of the amphibians were, as their name suggests, somewhat in between, preferring a more terrestrial lifestyle. The amniotes evolved from among their number. At first, therefore, the amniotes looked much like the amphibians with which they shared their world: all were fairly small, salamander-like creatures.[10] Like the amphibians, they scuttled and hid in the shell-hole stumps of lycopods, darting out to feed on cockroaches and silverfish, and avoiding the attentions of larger, nightmarish creatures that the abundance of oxygen had swelled into monsters. These amniotes dodged the stingers of scorpions the size of dogs; hid from millipedes as long and wide as magic carpets; and presumably quailed at the remorseless, spiky, tank-like tread of 2-metre eurypteryid sea scorpions that had left the ocean in pursuit of their rapidly evolving fishy prey.

For an amphibian, laying eggs in this Garden of Earthly Delights was supremely hazardous. To spawn in open water, like a modern frog or toad, was to provide an easy snack for any passing fish, or another amphibian. Amphibians had to evolve various ways to protect their offspring. Some stood guard over their spawning grounds. Others sought pools and puddles away from open water – in tree stumps, for example; or perhaps they deposited spawn in jellylike masses in vegetation hanging over water, so that, when the tadpoles hatched, they would drop straight in. Yet others would prolong the larval stage, hatching not as tadpoles, but as miniature adults, fully equipped to scuttle away from any threat. Others went the whole way and retained the eggs inside the mother, perhaps even nurturing them on maternal tissue, and giving birth to them, large and live.[11]

The amniotes went further. Their adaptation was not in where they laid their eggs, but in the eggs themselves. The helpless, hapless black dot of an embryo was surrounded not just by jelly, but by a series of membranes that would keep the dangerous world away for as long as possible.

One membrane was the amnion, a waterproof caul that provided the embryo with its own exclusive private pond and life-support system.[12] A yolk sac would keep it nourished. A further membrane, the allantois, collected and stored the embryo's waste. Surrounding them all was the chorion, and, surrounding *that*, the shell.

The shells of the earliest amniotes were soft and leathery, more like the shells of snake or crocodile eggs than the hard and crystalline eggs of birds.[13] Importantly, amniote eggs did not require the kind of energy-sapping and elaborate parental care that amphibians needed to devote to their spawn. Eggs could be laid, buried under leaf litter or inside a rotting log to keep them warm, and then forgotten about.

To begin with, the amniote egg was just another way for amphibians to improve the chances of their offspring against being eaten before they had even hatched. But these early egg-layers had also evolved a way to break free from water completely. The amniote egg was like a spacesuit for colonizing a new and hostile world – a world entirely away from water.

Within a few million years, true amniotes had evolved. No longer small and salamander-like, they were small and lizard-like. Animals such as *Hylonomus* and *Petrolacosaurus* looked similar, and did much the same things – foraging for insects and other small animals that could not escape their hungry jaws. Yet they were close to the lineages that eventually were to produce snakes, lizards, crocodiles, dinosaurs and birds. The destiny of *Archaeothyris*, however, lay elsewhere. This creature was a pelycosaur, a member of a group of reptiles whose descendants were to include mammals, including us.

The evolution of the amniote egg was the key to the success of vertebrates on land. The world of plants, too, responded to the challenge of aridity in its own way, with the evolution of seeds, in a number of superficially fernlike relatives of what were, eventually, to become conifers. These were the seed ferns.

Liverworts and mosses, the earliest plants on land, are like amphibians, in that their reproduction is totally dependent on water. Male plants produce sperm that swim through the sheen of water that always coats the leaves and stems of such water-loving plants, in search of female plants' eggs to fertilize. A fertilized egg develops into a plant that does not produce eggs or sperm, but tiny particles called spores. These spread through the environment, and, where they settle, germinate into more egg- and sperm-producing plants.

And so the cycle continues, with alternating generations of sex-cell-producing (gametophyte) and spore-producing (sporophyte) plants. Although spores are often resistant to desiccation, sperm and eggs are not, which is why mosses and liverworts are always tied to water.

In mosses and liverworts, gametophyte and sporophyte plants look rather similar. In ferns, though, the bias is very much towards the sporophyte. The ferns we see in woods and fields are all sporophytes. The spores are produced in long rows of capsules under the leaves. The gametophytes, by contrast, are small, tender and occult, and do not look very much like ferns at all: and, because they produce eggs, and sperm that move through a water film, they need damp places to survive. Likewise, the giant clubmosses and horsetails of the great coal forests.

In some of the ferns, though, the gametophytes shrank so much that they became hardly more than the sex cells they produced. So small, in fact, that the entire gametophyte generation was confined within the spores, which could be either male or female. In some species, female spores tended to remain attached to the plant, rather than being shed into the environment. Male spores would be carried to the female by the wind. The egg, once fertilized, became a seed, shielded in a tough, resistant rind, and would germinate only when it met the right conditions. The evolution of the seed, like the evolution of the amniote egg, allowed plants to break away from the tyranny of water.

The exuberant growth of the coal forests did not last. The debt became due with the slow northward movement of Pangaea. The southernmost regions, which had once lain over the South Pole and had been ice-covered for much of the later Carboniferous and early Permian, once more became free of ice. However, with the fusion of the northern and southern continents, there was no easy way for warm, equatorial water to circle the globe. There was too much land in the way.

There was an ocean, however, bursting with life. This was the Tethys, a large, tropical gulf fringed with reefs, in the eastern side of Pangaea, making the supercontinent as a whole look like a very large letter 'C'.

The lie of the land – with no easy way for equatorial water to circle the Earth – meant that the shores of the Tethys became very seasonal. Long, dry seasons were sharply punctuated by ferocious monsoon rains, similar to those that now drench India, but on a global scale.[14] This seasonal climate was too much for the lycopod-dominated rainforests, which required tropical moisture all year round. The rainforests shrank into isolated patches. The exception was South China, then an island continent to the far east of the Tethys, which retained lycopod forests: a land that time forgot.

The forests were replaced by a mixture of spore-producing tree ferns, seed ferns and smaller lycopods, generally adapted for a more seasonal climate in which much of the year was dry and very, very hot. Away from the coastlands, deserts spread.

The effect of the demise of the coal forests on the fortunes of amphibians and reptiles was stark.[15] Amphibians suffered, though reptiles managed to hang on, and adapted to the opportunities afforded by the drier climate.

Although many of the amphibians remained crocodile-like and lived close to water, a few took up the challenge of desert life, and looked very much more reptilian. One of these was *Diadectes*, a rhino-like animal that grew to 3 metres long, and was something of a pioneer. It was one of the very first tetrapods to embrace a radical new diet – vegetarianism. Until that point, all tetrapods had eaten insects, fish, or one another. Meat is hard to catch, but when caught, is quick and easy to digest. Plants, however, are constrained to stand and fight, and they do this with tough, fibrous tissues, in which each cell is armoured in a curtain wall of indigestible cellulose.

If plant matter cannot be broken down mechanically

– and the first tetrapods did not have especially effective grinding teeth – the plant matter has to be cropped and snipped and swallowed and slowly fermented, like compost, in a capacious gut by a range of bacteria, the niggardly nutrition released only very slowly. This is why herbivores tend to be large, slow-moving, and eat virtually all the time. *Diadectes* was joined by the first reptilian herbivores. Among them were massive, warty pareiasaurs, heirs of small, lizardy *Hylonomus*, like buffalo on steroids; and a variety of pelycosaurs such as *Edaphosaurus*, an altogether more elegant creature that sported a membrane on its back supported by greatly elongated vertebral spines.

These herbivores were preyed on by land-living amphibians such as *Eryops*, which looked like a bullfrog imagining itself as an alligator. Had it had wheels, it would have been an armoured personnel carrier. With teeth. Battling *Eryops* for the top spot were other sail-backed pelycosaurs such as *Dimetrodon*.

Unlike mammals and birds, reptiles and amphibians have no internal control of their body temperature. Torpid and defenceless in the cold, they need to warm themselves in the sun before they can become active. This created an opportunity for animals that could heat up and cool down more quickly than others. Pelycosaurs were among the first tetrapods to take active control of their metabolism. When they stood with their sails sideways on to the sun, *Edaphosaurus* or *Dimetrodon* could warm up much more quickly than reptiles without such a facility – and be the first to the feeding grounds. When the sails were placed edge on to the sun, they could shed heat more quickly, too. Pelycosaurs evolved another trick: unlike most reptiles, whose jaws have rows of identical, pointed teeth, pelycosaurs started to evolve teeth of different sizes, enabling them to process food more efficiently.

These adaptations – heat regulation, and the development of differently sized teeth – were signs of things to come.

A scion of the pelycosaur lineage was *Tetraceratops*,[16] which lived in the early Permian deserts of what is now Texas. Although very much like a pelycosaur, there are signs in its skull and teeth of changes to something wholly different, a new order of reptile in which some of the metabolic innovations of pelycosaurs were taken very much further. These creatures were called the therapsids.[17] Commonly referred to as 'mammal-like reptiles', they were the stock from which mammals were, eventually, to evolve. In the mid-Permian, however, all that still lay tens of millions of years in the future.

Therapsids differed from pelycosaurs and other reptiles in that they tended to hold their limbs upright and underneath the body, rather than sprawling out to the sides. They had a wide variety of interesting teeth, suited to their diets, and they were warm-blooded; that is, they could regulate their own metabolisms, independent of what the sun was doing. Therapsids dominated the dry, seasonal landscapes of Pangaea. They eclipsed the pelycosaurs, their kin, and all but drove the more land-loving amphibians back to the water.

For every ecological niche the mid- and late Permian had to offer, there was a therapsid that would slot neatly into it. Early therapsid herbivores included 2-tonne monsters such as *Moschops*. These were succeeded by the dicynodonts, arguably the most successful and ugliest tetrapods that ever walked the planet. These barrel-shaped creatures ranged in size from small dog to rhinoceros. Their heads were wide but their faces flattened, as if they spent their lives chasing parked cars. All their teeth had been replaced by a horny beak, except for a pair of very large, tusk-like upper canines. Although nominally herbivores, dicynodonts shovelled up anything and everything they came across and could stuff into their mouths. Some of the smaller ones could burrow. Both habits, as it turned out, would serve to protect them from the apocalypse to come.

Dicynodonts were stalked by ferocious predators – their therapsid cousins, the gorgonopsians. Like dicynodonts, they varied in size, from badgers to bears, but apart from their avoidance of parked cars, were otherwise very similar. Slouchy, rangy quadrupeds, they sported enormous upper canine teeth, comparable to those of sabre-toothed tigers. Other carnivorous therapsids included the cynodonts. These tended to be smaller than gorgonopsians, and the later ones were smaller still.

As the Permian wore on, cynodonts all but relegated themselves to the margins. They were small, sometimes nocturnal. They had large brains, and teeth that were fully differentiated into incisors, canines and molars. They had fur, and whiskers. They shared the margins of their world with the small, generally lizard-like descendants of *Petrolacosaurus* and *Hylonomus*.

At its greatest, Pangaea stretched almost from pole to pole. The union of the continents into a single landmass had drastic consequences for life both on land and in the oceans. On land, forms of life that were once endemic to particular continents mixed and mingled with others. Competition between natives and newcomers was fierce, and many kinds of animals died out.

Sea life was most abundant on the continental shelf, the part of the sea closest to land. When the continents merged, there was less continental shelf to go round. Competition for living space in the sea was, therefore, also intense.

The climate itself became more challenging. The interior of Pangaea was mainly dry, even if punctuated by annual monsoon inundations, and – with the northerly drift of the entire landmass – often very hot indeed. Although the cool southern regions of Pangaea were clothed in a seemingly endless scrub of a tree-fern called *Glossopteris*, plant life was not as luxuriant as it had been. Less plant life meant that there was less oxygen than there once was, so much so that by the end of the Permian, breathing at sea level would have been like trying to catch a breath in the Himalayas today. Terrestrial life was left gasping.

Worse was to come, for Armageddon was approaching. Near the end of the Permian, a plume of magma[18] that had been rising from deep within the Earth for millions of years met the crust above, and melted it.

In the late Permian, one would not have needed to

descend into the Earth to find Hell, because Hell had come to the surface. It lay in what is now China, where what had once been a lush landscape of rainforest was transformed into a cauldron of magma, oozing lava and a fume of noxious gases that increased the greenhouse effect, acidified the oceans and tore the ozone layer to shreds, bringing down the Earth's shield against ultraviolet radiation.

Life had not quite recovered from this disaster when, around 5 million years later, another struck. The China magma plume, as it turned out, was merely the *hors-d'oeuvre*. The *entrée* was an even larger magma plume which, rising from deep within the Earth, punctured the Earth's surface in what is now western Siberia.

The ground fractured. Lava oozing from myriad fissures eventually paved an area the size of what is now the continental United States, from the eastern seaboard all the way to the Rocky Mountain front, in black basalt thousands of metres thick. The ash, smoke and gas that accompanied it killed nearly all life on the planet. But not instantly: the torture was extended into half a million years of toxic agony.

First of the evil brew was carbon dioxide, enough to create a greenhouse effect that raised the average temperature of the Earth's surface by several degrees. Already starved of oxygen and labouring under searing heat, parts of Pangaea became completely uninhabitable.

The effect on the reefs fringing the Tethys Ocean was nothing short of catastrophic. The sun-loving algae living inside the jellylike polyps that made up the coral reefs were acutely sensitive to temperature. When the temperature of the sea rose, they deserted their homes, leaving the polyps to die.[19] The coral, bleached and dead, crumbled away.

Tabulate and rugose corals, the mainstay of reef ecosystems for tens of millions of years, were already in decline as a result of changing sea levels, but the Siberian event was the final straw.[20] Without the coral, the host of organisms that depended on them for a habitat died out, too.

But there was more. The volcanoes scorched the sky with acid. Sulphur dioxide frothed high into the atmosphere. There, sulphur dioxide helped form microscopic particles about which water vapour condensed to create clouds that reflected sunlight into space, cooling the Earth's surface, albeit temporarily. Amid the heat there were jags of bitter cold. However, when rained out onto the land, sulphur dioxide became an acid that stripped plant life from the ground, leached the soil, and burned forest trees to blackened stumps where they stood. Traces of hydrochloric and even hydrofluoric acid sharpened the pain. And before it was rained out, the hydrochloric acid damaged the ozone layer that protected the Earth from harmful ultraviolet rays.

In normal times, plankton in the sea and plants on land would have mopped up much of the carbon dioxide. But plant life was already under stress. So, rather than being absorbed by plants, the carbon dioxide was washed out by rain, increasing the rate of weathering.

Without plants to stabilize the soil, the weather washed it away, leaving bare rock. The sea became a thick soup, turbid not just with sediment but with the crouton carcasses of organisms – plants and animals – killed by the carnage on land. Decay bacteria got to work on the remains, using up what little oxygen was left. A corpse-flowering of toxic algae had much to digest, before that, too, withered. The bubbling acids in the water etched the shells of any sea creature they touched, dissolving them. Even if they survived the darkened, stagnant sea, the mineralized skeletons on which many sea creatures depended became thin and fragile, until they could no longer make shells at all.

And even more was to come. The mantle plume destabilized deposits of methane gas, hitherto frozen in ice beneath the Arctic Ocean. The gas fizzed to the surface of the sea with a thunder and spume that shot hundreds of metres into the atmosphere. Methane is a much more potent greenhouse gas than carbon dioxide. The greenhouse effect spiralled; the world broiled.

If that wasn't enough, every few thousand years the eruptions would send plumes of mercury vapour into the atmosphere,[21] to poison anything that had not already been asphyxiated, gassed, burned, boiled, broiled, fried or dissolved.

By the end, nineteen of every twenty species of animal in the sea, and more than seven out of every ten on land, had been driven to extinction. Among the dead were animals that have left no descendants or close relations.

The extinction killed the last of the trilobites, for example. These busy pill-bug-like creatures had scuttled over the sea floor and swum above it since early in the Cambrian. By the Permian, however, they had been in decline for a very long time, and few enough remained by the Permian that their final departure was voiced quietly, in a minor key.

Likewise the blastoids, a group of stalked echinoderms. Between the Cambrian and the Permian there were as many as twenty kinds of echinoderm, of which the blastoids were among the last to survive. The echinoderms that are still with us, by contrast, are in many cases so familiar that any beachcomber takes them for granted. Today, there are just five kinds: sea stars, brittlestars, sea cucumbers, sea urchins and feather stars.[22]

However, it could so very easily have been four. But for just two species in one genus of sea urchin that weathered the storm, the sea urchins, too, would have been consigned to oblivion. The survivors hung on, evolving and diversifying into all the sea urchins alive today. Although modern sea urchins are very diverse, from globular purple urchins to almost-flat sand dollars, the sea urchins of the Palaeozoic were more varied still. All modern sea urchins, however, draw from the limited genetic pool that was the legacy of the few that survived the cataclysm. But for the hardihood of these few witnesses to destruction, sea urchins would have been completely absent from modern seashores, and as remote and exotic to us as the blastoids.[23]

Virtually all shellfish perished, whether burned by acid, or drowned in decay amid an airless sea. Only a very few species survived. One was *Claraia*, a bivalve that looked rather like a scallop. Back in the Permian, and before it, the kings of the sea were creatures called brachiopods. These look superficially like bivalve molluscs, in that they have soft bodies enclosed in two shells like hands cupped in prayer, and make a living by sieving detritus from the water. The end-Permian extinction tipped the balance. Almost all brachiopods became extinct, and they are very minor players in the modern ocean ecosystem. The spoils went to *Claraia* and its descendants, explaining why it is bivalves such as cockles and mussels (scallops, too) that litter the seashore today, whereas brachiopods are generally only found as fossils. The end-Permian extinction had lasting consequences for the shape of life that resonate to this day.

On land, the lives of generations of amphibians and reptiles were swept away. The legions of lumbering, horned and warty pareiasaurs all vanished. The sail-backed pelycosaurs, likewise, did not survive the Permian. Neither did most of their relatives, the therapsids. The vast herds of dicynodonts that cropped horsetails and ferns on the Permian plains were almost entirely culled, along with the sabre-toothed gorgonopsians that had pursued them.

Amphibians were almost completely driven back to the water whence they had first emerged, back in the Devonian. All those that had forged a life on land, becoming more reptilian in life and habits, became extinct. The ancestor of all the amniotes had emerged from within this group of creatures, back in the early Carboniferous, making land life a much more viable proposition. Nothing like them survives today.

The gates of Hell, ajar in China, thrown extravagantly wide in Siberia, had sucked almost all life into the abyss. The land was turned into bare, silent desert; little plant life remained, clinging onto the wreckage of what was largely a dying planet. The ocean was all but dead. The reefs were gone, the sea floor clothed with a stinking carpet of slime. It was as if life had been catapulted back to the Precambrian.

But life would return. And when it did, it would be as the most colourful, riotous carnival of splendour the world had seen yet.

6

TRIASSIC PARK

Recovery from the disaster that was the close of the Permian period took tens of millions of years. The world, once teeming with life both at sea and on land, was relatively bare. Ripe, then, for opportunists – such as the altogether remarkable *Lystrosaurus*.

With the body of a pig, the uncompromising attitude towards food of a golden retriever, and the head of an electric can opener, *Lystrosaurus* was the animal equivalent of a rash of weeds on a bomb site. *Lystrosaurus* was a dicynodont, a member of that once large and varied group of creatures, the therapsids, that had dominated the land in the Permian. A habit of burrowing its way out of danger might have saved it from the apocalypse that claimed the lives of most of its fellows.

Success after its emergence was all about its go-anywhere, eat-anything attitude, and its skull, which was wider than it was long. Massive chewing muscles drove a lower jaw that had lost all teeth save for a sharp-edged, horny beak. The upper jaw, too, was reduced to a blade, all except for a pair of canine teeth elongated into tusks, either side of a flat face. The powerful head worked like a backhoe – digging, scraping, scything and shovelling anything it could find into its ceaselessly chomping maw.

Immediately after the extinction, and for millions of years more, land life was a near-monoculture of *Lystrosaurus*. They ranged in herds all over Pangaea, and were as happy in the occasional woodlands or wetlands as they were in the hot, dry deserts that were typical of the period. There were other animals, to be sure, but nine out of every ten was a *Lystrosaurus*: arguably the most successful land vertebrate that has ever lived.

So, what, besides *Lystrosaurus*, survived? The amphibian dalliance with a much more land-based life, in animals such as *Diadectes* and *Eryops*, did not last. Triassic amphibians were aquatic, similar in their habits and appearance to crocodiles. Some of them, however, were very large – and a few of the larger kinds survived until the middle of the Cretaceous, antique remnants of a vanished age, before they, too, finally became extinct. The prize, ultimately, went to the smaller forms. The first frog, *Triadobatrachus*, evolved in the Triassic.

Despite its global range, *Lystrosaurus* was much less common in the far northern and southern regions of Pangaea, particularly in the earliest Triassic. The polar regions of the early Triassic, though cooler than the fryingly torrid equatorial zone, were arid between the watercourses, where the giant amphibians still ruled.

The reptilian inheritors of the Triassic descended from those few small creatures that scuttled through the extinction, under the feet (and into the burrows) of *Lystrosaurus*.

Once in the Triassic, they diversified very quickly into a dazzling range of forms, a defiant riposte to the events that almost destroyed life beyond recovery.[1] Many of these newly minted reptiles took to the water.

Along with frogs, the turtles were a group of animals that first evolved in the Triassic, and likewise diversified in the water. Although the Triassic form *Proganochelys* looked like a modern land-living turtle, with a fully formed shell above and underneath, other Triassic turtles included *Odontochelys*, which had a fully formed shell on its belly (the plastron), but only a partial carapace above consisting of broad ribs;[2] the terrapin-sized *Pappochelys*, in which the carapace and plastron alike were yet to form fully;[3] and the metre-long *Eorhynchochelys*, which had neither plastron nor carapace, and combined a very unturtle-like long tail with a very turtle-like beak.[4] The Triassic was a golden age for turtles, almost-turtles and even mock turtles, in which they adopted a very wide variety of forms and modes of life.

Superficially similar to the turtles were the placodonts,[5] slow-moving marine reptiles with thick bodies, often armoured with a carapace, and tombstone-like teeth specialized for crushing the shells of molluscs. While placodonts grubbed in the ooze for shellfish, other reptiles – the nothosaurs, and the rather similar thalattosaurs and pachypleurosaurs – darted through the sparkling seas above in search of fish. These creatures were rangy, with long necks and tails, and limbs that served as flippers. Nothosaurs were related to the often much larger and even more aquatic plesiosaurs, which evolved much later. Nothosaurs, pachypleurosaurs and thalattosaurs, however, like placodonts, all lived and died in the Triassic.

Prowling the shallows and dipping for fish was *Tanystropheus*, a 6-metre-long creature whose neck was as long as, or longer than, its body and tail combined. More curiously still, the neck was very stiff, being made of only a dozen or so extremely long vertebrae. Of all the oddities in the reptilian carnival freak show that was the Triassic, *Tanystropheus* was one of the oddest.

But then there were the drepanosaurs.

These unlikely creatures spent much of their time hanging over water, suspended by their prehensile tails, each of which had a stiff claw at the end to act as a grappling hook. Thus suspended, they would swipe the water to catch fish, aided by hook-like claws on one of the digits on each forelimb, eventually spearing and swallowing them with their long, birdlike bills.[6]

Sharing the Triassic seas were the hupehsuchids,[7] a small group of aquatic reptiles with stumpy, flipperlike limbs and long, beaklike snouts. These strange creatures were related to that acme of aquatic reptiles – the ichthyosaurs. These animals, which also appeared in the Triassic, looked superficially like dolphins. They lived their entire lives at sea, and gave birth to live young, as whales do. Some of them grew to whale-like sizes. The Triassic *Shonisaurus*[8] grew to 21 metres in length, not only the largest ichthyosaur but the largest marine reptile known. Although ichthyosaurs lived until late in the Cretaceous, there were none to match them in their Triassic heyday.

On land, the monstrous, horned and warty pareiasaurs of the Permian had grazed their last: not so, their much smaller and rather distant cousins, the procolophonids. These small, squat and spiny creatures had broad skulls crammed with teeth suitable for milling vegetation or insects. No Triassic undergrowth of ferns and cycads was complete without one or more of these inconspicuous yet industrious creatures. To part a frond was to see one or more of them scuttling off into the shadows. In the Triassic, procolophonids were everywhere – by its end, they had all gone.

At the time, though, they might easily have been mistaken for the similarly spiny and lizard-like sphenodontians, which were, like the procolophonids, ubiquitous. Unlike the procolophonids, however, the sphenodontians survived, and still live – if only just – in the present day. The only remaining sphenodontian is the tuatara, now confined to a few small islets off New Zealand, the last of a lineage that stretches back almost a quarter of a billion years.

As with sphenodontians, so with the earliest true squamates – the ancestors of modern lizards and snakes. These, too, started out in the Triassic, with forms such as *Megachirella*.[9] Many small, early reptiles looked superficially like lizards, but *Megachirella* really was one.

Like the small amphibians of the Carboniferous, lizards had a tendency to lose their legs. This happened many times in lizard evolution. The culmination of this trend was the appearance of snakes, but that still lay in the future, in the Jurassic period, when the breakup of Pangaea led to an evolutionary flowering of both lizards and snakes.[10] Not that snakes lost their limbs all at once – some early forms kept their hindlimbs. The Cretaceous snake *Pachyrhachis*, which slithered off the southern shores of the Tethys, had tiny, vestigial hindlimbs.[11] Another form, *Najash*, had much sturdier hindlimbs, attached to the sacrum and fully usable, and lived on land.[12] As soon as they evolved, therefore, snakes diversified into a range of burrowers and swimmers.

Lystrosaurus – and one or two other, rarer dicynodonts that weathered the end of the Permian – continued to evolve and diversify, giving rise to a range of similar but much larger animals such as the cow-sized *Kannemeyeria*. These creatures browsed the plains alongside the rhynchosaurs. These looked rather similar to dicynodonts, with plump bodies and beaklike snouts, but were more closely related to the rulers of the Triassic – the archosaurs, or Ruling Reptiles.

Early archosaurs were not all small. One of the earliest was the enormous, and enormously terrifying, *Erythrosuchus*, a 5-metre monster that evolved to take advantage of the abundant mobile larder that was *Lystrosaurus*.

Today, the archosaurs are represented by two very different kinds of animal – crocodiles, and birds. In the Triassic, birds did not yet exist, but there was a bewildering range of animals that looked more or less like crocodiles.

Perhaps the closest were the phytosaurs, which would have easily been mistaken for crocodiles, except for their tendency to have nostrils at the top of the head, rather than at the end, to allow easy swimming underwater with only a minimal surface showing above it. Phytosaurs were carnivores, or, perhaps, fish-eaters. Their relatives, the aetosaurs, were vegetarians, protecting themselves with spiky and armoured carapaces, a premonition of the ankylosaurs that would evolve 100 million years later.

Aetosaurs would have had more to fear from the formidable rauisuchians, quadrupedal predators that grew up to 6 metres long, with deep, powerful skulls that looked uncannily like those of large meat-eating dinosaurs such as *Tyrannosaurus*. Although many alligators sprawl, they are also capable of a gait called a 'high walk' in which their limbs are held more tightly under the body. This is much more energetically efficient for land life. Rauisuchians walked like this, and so did many of their archosaurian relatives. A few, though, were bipeds, at least some of the time.

At sea, on land – and in the air. The Permian and Triassic saw several essays in flight by vertebrates, keen to pursue the insects which had taken to the medium back in the Carboniferous, and which also diversified strongly in the Triassic into a range of unusual forms. Various kinds of gliding reptiles chased dragonflies through the Permian and Triassic forests: creatures such as *Kuehnosaurus*, which looked and behaved much like the extant gliding lizard *Draco*. Another, more typically Triassic form – in that it was very strange, and nothing like anything seen before or since – was *Sharovipteryx*. This glided through the trees using a web of skin stretched between its greatly elongated hindlimbs.

It wasn't until the Triassic Period, however, that vertebrates started to fly properly, rather than simply glide from tree to tree. These aeronauts were the pterosaurs (once known as pterodactyls), which were archosaurs, and close cousins of dinosaurs.[13] Their wings were elastic membranes of muscle and skin stretched between hands and body on one massively elongated ring (fourth) finger – the word 'pterodactyl' means 'wing finger'. The first pterosaurs were small and flappy, rather like bats. And, like bats, they were also rather fluffy.

As pterosaurs evolved, they grew, until the last of their kind, at the end of the Cretaceous period, were as large as small aeroplanes, and barely flapped at all. Light in build, but with enormous wings, all they needed to do to take off was to spread their wings into a light breeze and physics would do the rest. Their success was abetted by a delicate construction, their skeletons modified into rigid, boxy airframes made of bones hollowed almost to paper-thinness. The largest pterosaurs were adapted for soaring on thermals in still air. These living sailplanes could turn incredibly tightly to take advantage of even the narrowest drainpipe of a thermal, even within the spans of their own wings, rising higher and higher until – at altitude – they'd break from the thermal and glide downwards to catch another one.[14] In this way they could travel a long way with almost no effort at all. Giant pterosaurs such as *Pteranodon* cruised the seas that opened as Pangaea broke apart, winging between the young and diverging continents.

Only the really big pterosaurs, like *Pteranodon*, the gigantic *Quetzalcoatlus* and the arguably even larger *Arambourgiana*, could have soared in this way. No amount of power could have flapped those huge wings without crumpling them. And pterosaurs didn't have the keeled breastbones of birds that anchor their powerful flight muscles (these are the breast muscles on your table bird). Only small pterosaurs had wings small enough for feasible, bat-like flapping.[15] The last and largest pterosaurs did not, in fact, fly much at all, but lumbered on the ground like vast mobile marquees, their huge heads capable of seeing eye to eye with a giraffe.

The break-up of Pangaea was an opportunity for snakes and lizards. For the pterosaurs cruising the currents of air far above, it was their undoing. Continental drift during the Jurassic and Cretaceous created a varied and stormy climate, very different from the more even temperatures of the Triassic. Even though the climate of Pangaea was often very challenging, the winds, outside the monsoon season, were light. The absence of ice at the poles, and the freedom of the ocean to circulate warmth to all latitudes, meant that the temperature gradient between poles and Equator was very shallow. When the climate turned breezier, however, these giant, delicate kites of creatures were hurled headlong, crashing to the ground like so many broken umbrellas, and smashing apart on impact.

Amid the riot of reptiles, a few – a very few – therapsids that weren't dicynodonts clung on. Early in the Triassic, dog-sized cynodonts such as *Cynognathus* and *Thrinaxodon* fulfilled the roles of small to mid-sized carnivores. As the ages wore on, the creatures in this lineage got smaller and smaller, and furrier and furrier, and, slipping almost unnoticed into neglected and nocturnal nooks, evolved into the mammals. But their time had not yet come.

Among the more bipedal archosaurs were the earliest dinosaurs, which emerged, late in the Triassic, from the riot of rauisuchians, rhynchocephalians and other more or less crocodile-like animals.

The roots of the dinosaurs and pterosaurs – the 'bird-line' archosaurs, as distinct from the lineage that led to crocodiles – lies in a group of Triassic creatures called the aphanosaurs, such as *Teleocrater* – a long and low-slung quadruped that looked rather crocodile-like but for the longer neck and small head.[16]

It would be hard to tell, looking at an animal such as *Teleocrater*, that its lineage had a marvellous and weighty destiny to look forward to, whereas almost all their archosaur kin would perish. There was a clue, however, in its bones. Aphanosaurs had a slightly higher growth rate than many other archosaurs, and were just a little more active and aware of their world.

Closer still to the dinosaurs were the silesaurs. These were more slender and graceful than aphanosaurs, with long tails and long necks: but they still had all four of their feet on the ground.[17] By the end of the Triassic, all the aphanosaurs, and all the silesaurs, had gone. Their closest relatives, the dinosaurs, however, embraced the two-legged stance as a way of life, rather than doing so occasionally. They based their entire anatomy around it. And they were to inherit the Earth.

The dinosaurs began quietly, in the warm, humid interior of Gondwana, far away from the storm-wracked coasts of the Tethys, and the alien heat of the deserts on either side. Although they had already started to diversify into the carnivorous theropods and vegetarian sauropods familiar from their later history, the dinosaurs were a relatively small sideshow in the Triassic carnival of dicynodonts, rhynchosaurs, rauisuchians, aetosaurs, phytosaurs and giant amphibians.

But as some of the larger herbivores – the dicynodonts and rhynchosaurs – began to decline, the herbivorous dinosaurs slipped into their places. The dinosaurs moved, too, into more northern regions, and, eventually, to the deserts toward the Equator that had previously been closed to them. Even then, they were still minor players in a much larger drama of crocodile-line archosaurs. Theropods such as *Coelophysis* and *Eoraptor*[18] were small, darting opportunists, very far from the monsters of the Jurassic and Cretaceous. Rauisuchians still ruled the roost on land; giant amphibians in rivers and lakes; and a wealth of other reptiles in the sea. The sauropods and their kin, such as *Plateosaurus*, were large, but not the ostentatiously huge land-whales such as *Brachiosaurus* or *Diplodocus* they were to become. Towards the end of the Triassic, there were no obvious signs that destiny would favour the dinosaurs more than any other reptilian group. Dinosaurs took up the middle seats in the Triassic reptilian orchestra, behind the star soloists, and stayed there for 30 million years.

And still, as always, beneath it all, the Earth moved. Pangaea, the supercontinent forged over hundreds of millions of years from the fragments of Rodinia, was itself starting to break apart.

It started along a weak point, a seam in the crust, where other such dramas had come and gone. Long before Pangaea, it marked the line where the Appalachians, running parallel with the eastern seaboard of North America, had formed from the collision of two continental plates back in the Ordovician, 480 million years ago, squeezing an earlier ocean out of existence.

Late in the Triassic, the crust started to pull itself apart, more or less along the same line, to create what would become a new ocean – the Atlantic. A great rift valley formed, a widening gash in the Earth, from the Carolinas in the south to the Bay of Fundy in the north. As it widened, sediments on either side slumped into the gap, creating an ever-changing patchwork of rivers and lakes, full of life, but with volcanoes looming on every side.

The time came when the crust was stretched so thinly that the monster lurking beneath could be unleashed. Around 201 million years ago, a pustule of magma burst onto the surface, covering eastern North America and the then-adjacent regions of North Africa in basalt, and releasing carbon dioxide, ash, smoke and the now-familiar cocktail of noxious gases. Global temperatures, already high, shot up to peaks still more inimical to life. It was as if the Earth, smarting from its failure to extinguish all life 50 million years earlier, had come back for another try.

This crisis lasted for 600,000 years.

By the end, the sea flooded into the rift, the beginnings of what would become the Atlantic Ocean. But many of the animals that would have sliced through the new-made seas were no more: the thalattosaurs, pachypleurosaurs, nothosaurs, hupehsuchids and placodonts were gone.

Ichthyosaurs survived, along with a scion of the nothosaurs, the plesiosaurs. On land, the dicynodonts and procolophonids, the rauisuchians and rhynchosaurs, the silesaurs, the bizarre *Sharovipteryx*, *Tanystropheus* and drepanosaurs – were all swept away. The great Triassic circus had left town, leaving a ragged band of survivors.

The variety of crocodile-like animals was whittled down to the lineage that gave rise to the crocodiles we see today. The giant amphibians also survived, if barely, along with pterosaurs, a very few mammals and their mammal-like cynodont therapsid relatives, newly emerged sphenodontians, turtles, frogs and lizards – and the dinosaurs.

That the dinosaurs survived, when so many similar crocodile-like creatures did not, remains a mystery. It might simply have been a matter of luck. After the Permian, it was *Lystrosaurus* that had won the lottery of life. But now it was the dinosaurs that would rise and diversify to fill the brand new world that had opened up.

7

DINOSAURS *in* FLIGHT

Dinosaurs had always been built to fly. It started with their commitment to bipedalism, which had always been rather greater than that of their many crocodile-like relations.[1]

Most habitually four-footed creatures have a centre of mass in the chest region. It takes a lot of energy for them to lever themselves upwards onto their hindlimbs. This makes it hard for them to stand comfortably upright for any length of time. In dinosaurs, by contrast, the centre of mass was over the hips. A relatively short body forward of the hips was counterbalanced by a long, stiff tail behind. With the hips as a fulcrum, dinosaurs could stand on their hindlimbs without effort. Rather than the stumpy, robust limbs of most amniotes, dinosaurs could grow their hindlimbs long and thin. Legs are easier to move if they are more slender towards the ends. The easier the legs are to move, the easier it is to run fast. The forelimbs, no longer required for running, were reduced, the hands left free for other activities, such as grasping prey items, or climbing.

Constructed as a long lever, balanced on long legs, dinosaurs had a system of coordination that monitored their own posture constantly. Their brains and nervous systems were as pin-sharp as any animal that ever existed. All this meant that dinosaurs could not only stand, but run, strut, pivot and pirouette with a poise and grace the like of which the Earth had not seen before. It was to prove a winning formula.

The dinosaurs swept all before them. By the end of the Triassic, they had diversified to fill every ecological niche on land, much as the therapsids had in the Permian – but with consummate elegance. Dinosaurian carnivores of all sizes preyed on dinosaurian herbivores, whose defence was either to grow to great size, or clothe themselves in armour so thick they resembled tanks. In the sauropods, dinosaurs reverted to being quadrupeds and became the largest land animals ever to have lived, some measuring more than 50 metres long and, in *Argentinosaurus*,[2] weighing more than 70 tonnes.

And yet even they did not escape predation entirely. They were preyed on by gigantic carnivores: land sharks such as *Carcharodontosaurus* and *Giganotosaurus*,[3] culminating – in the very last days of the dinosaurs – in *Tyrannosaurus rex*.

In this single creature the potential of the dinosaurs' unique construction was taken to its greatest extreme. The hindlimbs of this 5-tonne monster were twin columns of sinew and muscle in which the speed and grace of its ancestors were traded for prodigious power and almost unstoppable force.[4] Balanced on its mighty hips by a long tail, the body was relatively short, the forelimbs reduced to mere vestiges, the mass concentrated in the powerful neck muscles and deep jaws. The jaws were full of teeth, each one the size, shape and consistency of a banana, if bananas were harder than steel. These were capable of bone-crushing force,[5] piercing the armour of slow but otherwise well-defended bus-sized herbivores such as ankylosaurs, and the many-horned *Triceratops*. *Tyrannosaurus* and its relatives tore bloody chunks from their prey and swallowed them whole – meat, bone, armour and all.[6]

But dinosaurs also excelled at being small. Some were so small they could have danced in the palm of your hand. *Microraptor*, for example, was the size of a crow, and weighed no more than a kilogram; the peculiar, bat-like *Yi*, diminutive in name as well as size, weighed less than half that.

The range of size in therapsids had been from large elephant down to small terrier, but dinosaurs exceeded even these extremes. How did dinosaurs get to be so very large – and so very small?

It began with the way they breathed.

There had been a rupture, deep in amniote history. In the mammals – the last surviving therapsids, Triassic throwbacks still gamely hanging on in the dinosaurs' shadow – ventilation was a matter of breathing in, and breathing out again. Considered objectively, this is an inefficient way to get oxygen into the body and carbon dioxide out. Energy is wasted drawing fresh air in through the mouth and nose, and down into the lungs, where oxygen is absorbed into the blood vessels surrounding the lungs. But the same blood vessels must shed waste carbon dioxide into the same spaces, which must be exhaled through the same holes by which fresh air came in. This means that it is very hard to clear all the stale air at once, or to fill every corner and crevice with fresh air in a single inspiration.

The other amniotes – dinosaurs, lizards and others – also breathed in and out through the same holes, but what happened between inspiration and expiration was rather different. They evolved a one-way system for air handling, which made breathing very efficient. Air entered the lungs, but did not immediately come out again. Instead, the air was shunted, guided by one-way valves, through an extensive system of air sacs throughout the body. Although seen in some lizards to this day,[7] it was the dinosaurs that elaborated this system to its highest degree. Air spaces – ultimately, extensions of the lungs – surrounded the internal organs, and even penetrated the bones.[8] Dinosaurs were full of air.

This system of air handling was as elegant as it was necessary. With powerful nervous systems and active lives that demanded the acquisition and expenditure of large amounts of energy, dinosaurs ran hot. Such energetic activity required the most efficient transport of air to oxygen-hungry tissues that might be contrived. This turnover of energy created a great deal of excess heat. Air sacs are good way to shed it. And this was the secret of the enormous size some dinosaurs achieved – they were air-cooled.

If a body grows but retains its shape, its volume will grow much faster than its surface area.[9] This means that as a body gets larger, there is much more of it on the inside relative to the outside. This can become a problem for acquiring the food, water and oxygen that a body requires – as well as voiding waste products, and the heat generated by digesting food, and simply living. This is because the area available for getting things in and out shrinks relative to the volume of tissues that must be so served.

Most creatures are microscopic, so none of this is a problem, but for anything much bigger than a punctuation mark, it becomes an issue. This is resolved, first, by evolving specialized systems of transport, such as blood vessels, lungs and so on; and, second, by changing shape, creating extended or convoluted systems that act as radiators, from the sails of pelycosaurs and the ears of elephants to the inner complexities of the lungs, which serve the important function of dissipating excess heat, in addition to gas exchange.[10]

The mammals, when they were eventually liberated from a world dominated by dinosaurs and able to grow to anything larger than a badger, solved this insulation problem by shedding hair as they grew, and by sweating. Sweat secretes water onto the skin surface, and as this evaporates, the energy required to transform liquid sweat into vapour is shed by tiny blood vessels just under the skin, creating a cooling effect. But exhaled air from the lungs also accounts for heat loss – which is why some of the furrier mammals pant, exposing a long, wet tongue to the evaporative relief of the air. The very largest land mammal was *Paraceratherium*, a tall, spindly and hornless relative of the rhinoceroses, which lived around 30 million years ago, long after the dinosaurs had vanished. It grew to around 4 metres at the shoulder, and weighed up to 20 tonnes.

But the largest dinosaurs were much, much larger than this. The surface area of a gigantic sauropod such as the 70-tonne, 30-metre-long *Argentinosaurus*, among the very largest land animals ever to have existed, was minute compared with its volume. Even changes in shape, such as extending the neck and tail, were not enough to shed all the heat generated by its capacious insides.

Although sauropods were very large, it's a rule of thumb that large animals have more relaxed metabolic rates than smaller ones, so generally run a bit cooler. Warming a dinosaur that size in the sun would have taken a long, long time – but cooling it would have taken just as long, so a very large dinosaur, once warmed, could have maintained a fairly constant body temperature simply by being very large.[11]

It was, however, the heritage of the dinosaurs that saved them, and allowed them to grow so big. Because their lungs, already voluminous, were extended into a system of air sacs that ramified throughout the body, these animals were less massive than they looked. Air sacs in the bones also kept the skeleton light. The skeletons of the largest dinosaurs were triumphs of biological engineering, the bones reduced to a series of hollow, weight-bearing struts, with as few non-weight-bearing parts as possible.

But the key was the fact that the internal system of air sacs did more than conduct heat from the lungs. It took heat from the internal organs directly, without first having to transport it around the body, via the blood, then to the lungs, dissipating some of it on the way, compounding the problem. A sizeable beneficiary was the liver, which generated a lot of heat, and which, in a large dinosaur, was the size of a car. The air-cooled internal workings of dinosaurs were more efficient than the liquid-cooled mammalian version.[12] This allowed dinosaurs to become much larger than mammals ever could, without boiling themselves alive.

Argentinosaurus was less a cumbrous behemoth, than a light-footed, quadrupedal, flightless . . . bird. For it is the birds, the inheritors of dinosaurs, that have the same lightweight structure, the same fast-running metabolism, and the same system of air-cooling. All of which are enormously advantageous for flight, an activity that demands a light airframe.

Flight is also associated with feathers. A coat of plumage was a feature of dinosaurs from very early on in their history. At first the feathers were more like hairs, a feature shared with pterosaurs, the first group of vertebrates that learned to fly, back in the Triassic – and close relatives of the dinosaurs.[13] Even without flight, a coat of feathers offered essential insulation for a small animal that generated a lot of heat. The problem faced by small, active dinosaurs was the opposite of that which challenged the very large ones – keeping all that expensive heat from dissipating into the environment.[14] But such simple feathers soon developed vanes, barbs and colour.[15] Animals as intelligent and active as dinosaurs had busy social lives, in which social display played an important part.

Another key to the success of the dinosaurs was laying eggs. Although vertebrates in general have always laid eggs – a habit that allowed the final conquest of the land by the first amniotes – many vertebrates have reverted to the ancestral habit, found in the earliest jawed vertebrates, of bearing live young. It is all a matter of finding a strategy that protects the offspring, without incurring too onerous a cost on the parent. Mammals started by laying eggs. Almost all of them became live-bearers, but at terrible cost. Live-bearing demands vast expenditures of energy, and this sets limits on the sizes that mammals can achieve on land.[16] It also limits the number of offspring they can produce at once.[17]

No dinosaur, however, ever nurtured its offspring in this way. All dinosaurs laid eggs – as do all archosaurs. Being intelligent, active creatures, dinosaurs maximized the success of their offspring by incubating the eggs in nests and looking after the young after hatching. Many dinosaurs, particularly the more gregarious herbivores such as the sauropods, and the smaller, more bipedal hadrosaurs that largely replaced the sauropods in the Cretaceous, made their nests in communal rookeries that dominated the landscape, stretching from horizon to horizon. Female dinosaurs drew on the insides of their own bones to provide enough calcium for their eggs, a habit that birds have retained.[18]

It was a sacrifice worth making in view of the advantages offered by egg laying. The amniote egg is one of the masterpieces of evolution. It consists not just of an embryo, but a complete life-support capsule. The egg contains enough food to bring an animal to hatching, as well as a waste-disposal system to ensure that this self-contained biosphere won't be poisoned. The act of laying an egg meant that a dinosaur was free from the trouble and expense of nurturing young inside her own body.

Some dinosaurs did expend energy caring for their offspring after hatching – but they were not tied to this obligation. Some buried their eggs in a warm hole or midden and left the young to take their chances. The energy otherwise expended on reproduction and parenting a small number of offspring could have been spent elsewhere – for example, by laying a much larger number of eggs than any amount of internal nurture would have allowed. And, of course, by growing. Dinosaurs grew rapidly. Sauropods needed to grow as rapidly as possible, until they were too large for carnivores to tackle. Carnivores had to grow rapidly in response. *Tyrannosaurus rex*, for example, achieved its adult mass of 5 tonnes in less than twenty years, growing by up to 2 kilos a day – a growth rate far faster than its smaller relatives.[19]

Dinosaurs and their immediate relatives spent millions of years accumulating everything they needed for flight: feathers; a fast-running metabolism; efficient air-cooling to keep it under control; a lightweight airframe; and a singular devotion to egg laying.[20] Some dinosaurs used some of these adaptations to do very un-birdlike things, such as grow to a size that no land animal has yet surpassed. Eventually, though, dinosaurs were cleared for take-off. How, then, did dinosaurs take that final step, and get airborne?

It began in the Jurassic period, when a lineage of already small carnivorous dinosaurs evolved to become even smaller. The smaller they became, the more feathered their skins, as small animals with fast metabolisms need to keep themselves warm. These animals sometimes lived in the trees – all the better to escape the attention of their larger brethren. Some discovered how to use their feathery wings to stay aloft for longer – and so became the birds.

There is nothing magic about an aerofoil, like a wing. It is shaped in such a way that it disrupts the air through which it moves, making some parcels of air move extremely fast, while others rest in the stillness of eddies and swirls. The net result of all these changes in speed is an upward force on the wing. This force increases in proportion to the speed at which the wing is moving. This force is called 'lift'.

There are two ways to become airborne.

The first is from ground level, or from water. The would-be aeronaut runs as fast as it can into the wind, flapping its wings as hard as it can. Take-off is theoretically possible even if the wings are held stiffly horizontal, but no flying animal is that powerful a runner. Flapping, though, alters the velocity distribution of air moving around the wing to increase lift further, making the improbable possible.[21]

The other way to get airborne is to perch in a high place and fall out of it, letting acceleration due to gravity do the work. It is even easier if one can jump into a thermal – a column of warm air rising from the ground – for extra buoyancy.

The best flyers are very small, microscopic even, and go wherever the wind takes them. Most living organisms are small and have voyaged like this for time immemorial, whether they were spores from the earliest land plants carried on an Ordovician breeze; viruses sneezed from a tyrannosaur's nostrils; bacteria sloughed from its skin; spiders borne on floating strands of silk; tiny insects – all were and are a large, and largely unconsidered, aerial plankton that floats from just above the ground to the edge of space. A very small organism, or a spore, or pollen grain, needs no special adaptations such as wings to get airborne, when it can be carried many kilometres by the lightest gust.

And that's the trouble. Aerial plankton is subject to the whims of the wind, and cannot control where it goes. For very small flyers, intent on imposing some direction on their lives, wings are needed. But for something as small as a mote, the molecules of air seem appreciably larger than they do to something as large as, say, a bee, or a fly. To a speck of dust, the air is viscous, like water, or syrup, so flying is more like swimming. The wings of the smallest winged insects are more like bristles than aerofoils and work like oars, rowing them through the air.

For a creature large enough for the pull of gravity to become more significant than the movement of air molecules, the first stage of flight is merely a kind of controlled falling. Such is parachuting. Those parachutists that manage to travel horizontally further than they fall vertically are known as 'gliders' – but it is a controlled kind of falling, nonetheless.[22]

Animals have discovered this means of locomotion many times, from the so-called 'flying' snake that spreads its body out to make a kind of single wing, and the 'flying' frogs with enormous, parachute-like feet, to many, many kinds of gliding lizard-like reptiles, either extant, or known from the fossil record, with skin stretched out each side on enormously elongated ribs, or even bones in the skin itself. They have been doing it since at least the Permian. Many small mammals are accomplished parachutists, from the sugar gliders of South East Asia to a range of 'flying' squirrels, parachuting or gliding using folds of skin stretched between fore- and hind feet. Mammals learned to glide almost as soon they evolved. One of the most ancient groups of mammals, the haramiyids, took to the air in the Jurassic period[23] possibly even before the earliest known bird, *Archaeopteryx*.

It could be no coincidence that all these gliding animals live or lived in trees – and that the parachuting habit has evolved many times independently.[24] After all, to any creature fond of climbing trees, natural selection exerts a relentless toll on any animal that falls out. Any animal with any adaptation that minimizes its impact, allowing it to die another day, will have natural selection in its favour.[25]

It was only the smaller dinosaurs that had any prospect of becoming airborne, for, as we've seen, the laws of physics show that as one gets larger, the power requirements of flight scale up, too. Only small flyers can flap. Larger ones can only soar.

Dinosaurs used a combination of routes into the air – running and flapping, and gliding and falling. In any case, they became airborne by accident. Their feathery wings were in place long before flying was even an option. Many dinosaurs sported tufts of feathers or quills, and had done so for a long time.

But it fell to a lineage of small, carnivorous dinosaurs to develop full-scale plumage. Although these creatures were bird-like in many ways – they folded their arms as birds fold their wings,[26] they incubated their eggs as birds do,[27] and so on – some of them were too large, physically, to have become airborne.[28] And yet many had feathers, which they used as thermal insulation, for sexual display, as camouflage to avoid predators, or a combination of all these and perhaps other things too.

The first flights were no more than short hops, and could have started either from the ground or from slightly higher up. The wings of the first dinosaurs to take to the air worked just well enough, and no more, to get them into low branches to roost at night. The chicks, being even smaller, might have got further, using their stubby wings to help them run up steep slopes, or tree trunks.[29] And once up in the branches, what then? A dinosaur with even the most rudimentary wings, especially if small, jumped out, using its wings to slow the descent, with the occasional flap to give lift. *Archaeopteryx*, the iconic 'first bird', had fully feathered wings but lacked the deep keel on the breastbone that modern birds have to anchor the flight muscles. *Archaeopteryx*, then, might not have been a very powerful flyer, but would have been capable of flying short distances between branches, or up to low branches above the ground.

Archaeopteryx lived at the end of the Jurassic period, and was just one of a very varied flock of dinosaurs experimenting with flight. Some of the earliest flying dinosaurs were biplanes, with flight feathers on their legs as well as their wings. The most famous was the tiny dinosaur *Microraptor*, from China, which was a member of a group of dinosaurs called dromaeosaurs.[30] The dromaeosaurs were close cousins of *Archaeopteryx*, along with another group of small, intelligent bipeds, the troödontids. And, just like birds and dromaeosaurs, troödontids were experimenting with feathers, and, perhaps, a degree of flight. One troödontid, *Anchiornis*, had feathers on its arms and legs – *Microraptor*-style – and lived in the Jurassic period[31] before the appearance of *Archaeopteryx*.

One of the strangest experiments in flight was made by another small group of dinosaurs closely related to dromaeosaurs, troödontids and birds. These sparrow-to-starling-sized creatures almost certainly lived in trees. Although they were feathered – one of them, *Epidexipteryx*, had long, ribbon-like tail plumes[32] – their wings were webs of naked skin, like bats.[33] These creatures, the scansoriopterygids, were a short-lived dinosaurian experiment in batlike flight that had sparked into life, sputtered and died before even the first bird hatched, or the first bat weaned.

Another feature in the evolution of flight is how often animals contrive to lose it.[34]

Birds seem to waste no opportunity in giving up flight as soon as they are able. Not all birds are very good at flight in the first place. At least two entire orders of birds gave up flight long ago. One group is the ratites, such as ostriches, emus, cassowaries and kiwis, and their extinct relatives the moas of New Zealand and the *Aepyornis* or elephant bird of Madagascar, both of which were driven to extinction not long after human beings first made landfall there. The other is the penguins, which turned their wings into flippers for flying underwater. Both groups are extremely ancient. Other birds became flightless when, having arrived on isolated islands free from ground-living predators, they found that they could take it easy – examples include the flightless cormorant of the Galápagos Islands, the kakapo (a kind of parrot) from New Zealand, and the dodo (an outsize pigeon) of Mauritius.

However, there were several other groups, unrelated to ratites, all of which became extinct millions of years before human beings appeared. In the late Cretaceous, a primitive bird called *Ichthyornis*, which looked like a seagull with teeth,[35] flitted along the shores of a seaway that once bisected North America from north to south, as pterosaurs such as *Pteranodon* soared above. It was accompanied by *Hesperornis*, a large bird more than a metre long but with virtually no wings at all, and which – like penguins – probably lived as a diver after fish. Another Cretaceous bird, hen-sized *Patagopteryx*, which lived in Argentina at about the time that *Ichthyornis* and *Hesperornis* were cruising the beaches of ancient Nebraska, also seems to have given up flight. A group of dinosaurs known as alvarezsaurids consists of a clutch of very small, feathered creatures with long legs, but wings reduced to tiny stumps, each bearing a large claw. When they were first described by scientists, they too were thought to have been flightless birds.[36]

Flight is an expensive habit. Even though all the prerequisites for flight existed in the dinosaur ground plan almost from the beginning, flying was and is immensely demanding, so it is not surprising that many flyers gave it up when the opportunity arose. The smaller and more flight-capable members of the dromaeosaurs and troödontids tended to be earlier examples of their families: their descendants were larger, and more grounded. The later dromaeosaurids and troödontids were the dragons that fell to earth.

Birds became flightless even before they became birds.

Not that many did not continue to accept the challenge. The Cretaceous skies rapidly became filled with the tweets, squawks and trills of innumerable birds. Many belonged to the enantiornithines, a group of birds very similar to modern ones, except that they retained teeth, and claws on their wings. But birds of modern aspect started to appear well before the end of the Cretaceous. The late Cretaceous shorebird *Asteriornis*, for example, was a cousin to the group of birds that would eventually turn into ducks, geese and chickens.[37]

The Earth continued to change. By the end of the Cretaceous, Pangaea had broken up into the land masses we'd recognize, more or less, today. This led to the evolution of distinct kinds of dinosaurs in different places. A group of theropods called abelisaurs was generally found only in the southern continents, whereas ceratopsians, such as *Triceratops*, were almost always found in western North America and eastern Asia – regions that were joined to each other, but separated from other landmasses.[38]

The isolation of dinosaurs on islands created strange Alice-in-Wonderland menageries. In the Jurassic, for example, Europe was an archipelago of tropical islands, much like Indonesia today, with its own unique fauna of miniature sauropods such as *Europasaurus*, each one no more than six metres long.[39] Madagascar, then as now, was a haven for the exotic. In the Cretaceous, many ecological niches there, even vegetarianism, were occupied by crocodiles.[40]

The Cretaceous also saw the appearance of flowering plants.[41] Flowering plants started small, and, like tetrapods, close to water, clothing riverbanks with the white, waxy blooms of water lilies that stood out starkly against the green wall of conifers.

Plants had long protected their embryos inside seeds, but flowering plants added further layers of protection. As in all plants, a male cell fertilized a female cell to create an embryo. But flowering plants added two other female cells which, both fertilized by another sperm, in a *ménage à trois*, created a tissue called the endosperm – tissue on which the young embryo could feed. The whole structure was enclosed in a further protective coat, which became the fruit. Before the fruit, there was the flower – coloured and scented to attract pollinators. The fruit, too, could be coloured and scented, to encourage animals to eat it and so disperse the seeds within, through their faeces.

Simple land plants, such as mosses, had been tempting animals to aid their fertilization for millions of years,[42] probably since the earliest colonization of the land. Such efforts were crepuscular and occult, and nothing like the spectacular first flush of flowering plants, which occurred alongside an explosive burst of evolution of pollinators such as ants, bees, wasps and beetles – creatures that dominate the Earth today in numbers of species. The relationship between flowering plants and their pollinators is subtle, many-faceted and complex – and only came into being when the age of the dinosaurs was at its height.

The world of the dinosaurs seemed as if it would never end. Indeed, it might have gone on indefinitely, despite the eructation of a magma plume in India at the end of the Cretaceous. Other than that, the Jurassic and Cretaceous were times during which the Earth seemed as though deep in slumber. The crisis that ended the Cretaceous, in contrast, was swift, brutal – and came from the sky.

One only has to look at the Moon's face to see that it bears the scars of collision. Most solid surfaces in the solar system are pocked with craters, from the microscopic to the gigantic. Even the tiniest fly-speck of an asteroid is peppered, crater upon crater, with the impacts of even smaller missiles. Only those bodies that constantly remodel their surfaces manage to erase this evidence.[43]

The Earth, too, has been struck by bodies from space many times, but craters that survive are rare. Those few impacting bodies that do not burn up in the dense atmosphere leave little in the way of scars, for these are soon worn away by wind, weather, water and, of course, the activities of living things. Worms burrow through the crater walls, undermining them. Roots crack them, crumbling them to dust. Seas fill them up, sediments bury them, life invades them, until they might never have been there at all.

But it only takes one. The impact of an asteroid some 66 million years ago brought the world of the dinosaurs to a sudden end.

Like all overnight sensations, the impact was a long time in the preparation. The dinosaurs' card had been marked long before. Around 160 million years ago, in the late Jurassic, a collision in the distant asteroid belt produced the 40-kilometre-diameter asteroid now known as Baptistina, along with a magazine of more than 1,000 fragments, each more than a kilometre across, some much larger. These harbingers of doom dispersed into the inner solar system.[44]

Around 100 million years later, one of them hit the Earth. Dive-bombing steeply from the north-eastern sky,[45] the body, which might have been as large as 50 kilometres across, hit the coast of what is now the Yucatan Peninsula in Mexico at 20 kilometres per second, penetrating and melting the crust. A blinding flash, followed by a 1,000-kilometre-per-hour gale carrying a noise beyond imagining, destroyed all life throughout the Caribbean region and much of North America, before the whole world was pelted with firebombs on a furnace-breath wind that turned trees into torches. Tsunamis pulled water all around the Gulf of Mexico well out to sea before a 50-metre wave crashed back to shore, surging more than 100 kilometres inland.

The impactor pierced sediments rich in anhydrite, left over from an ancient sea floor. Anhydrite is a form of calcium sulphate. The impact instantly converted it into sulphur dioxide gas. In the stratosphere, this gas created clouds. These – and the dust – shut out the Sun, plunging the world into a winter that lasted for years. By the time the Sun rose clear, the sulphur dioxide had been washed out as acid rain of stinging intensity, scarring the remaining plants and dissolving all reefs.

By then, all the non-flying dinosaurs were gone. The

last pterosaurs had been blown out of the sky. In the seas, the magnificent plesiosaurs – successors to the Triassic nothosaurs – perished, along with the mosasaurs, fearsome ocean-going monitor lizards.[46] The great ammonites, shelled relations of squid and octopus that cruised the seas in coiled shells, some as large as truck tyres, were eliminated, bringing to an end a pedigree that had started in the Cambrian.

The resultant crater was 160 kilometres across.

But once again life recovered. Although three-quarters of all species had been driven to extinction, life soon returned to ground zero. Within 30,000 years, the undersea site was already inhabited by plankton,[47] whose chalky skeletons, raining down on the sea floor, buried the remains of the impact crater.

The inheritors were those distant descendants of the therapsids, which, like the dinosaurs, had evolved a fast-running metabolism, but used it in an entirely different way. These were the mammals, and, after remaining in the shadows since the Triassic, they finally emerged into the light.

8

THOSE MAGNIFICENT MAMMALS

Once upon a time, back in the Devonian period, there was a pair of bones inside an armoured fish, in the back of the head, one on each side. The fish paid them no mind. The fish was, after all, busy, swishing sand into the dead-eyed stare of the giant sea scorpion that was chasing it.

The bones, however, continued to play their part. They were a pair of struts that braced the brain, in its case of soft cartilage, against the bony armour on the outside, just above the first pair of gill slits.

Jaws evolved when two other supports – the struts of cartilage that separated the mouth from the first gill slits – folded up on themselves in the middle, with the hinges pointing backwards. These jaw joints piled into the first pair of gill slits, reducing them to a pair of tiny holes. These were the spiracles, each one just above the jaw hinge on each side. The brain-bracing struts then found themselves doing triple duty. They were structural beams, as before. But they also, at one end, anchored muscles that opened and closed the spiracles. At the other, they were closely pressed to the paired holes in the braincase that led to the inner ears, one on each side.

The inner ears were tiny, fragile structures without which

the fish would become lost, disoriented, and would not know which way was up. They comprised labyrinths of fluid-filled tubes, each one a mirror image of the other. Movement in the fluid disturbed globs of mineral-enriched putty-like matter attached to hairs, which were attached at their other ends to nerve cells. Movement in the environment led to movement of the fluid, which disturbed the globs of matter, which tugged at the hairs, which fired nerve-triggered impulses to the brain – and, instantly, the fish knew where it was: swimming fast through the water with the snapping claws of a voracious sea scorpion close behind.

This same system of channels was sensitive to vibration in the water: again, through a system of microscopic hair cells, like the strings of a harp. Vibration would pluck the strings, each tuned to its own note, and the fish heard the ominous rumble of its pursuer. And the pair of ever-present, hard-working bracing struts, one on each side, conducted those vibrations from the outside all the way into the inner ear.

In the earliest tetrapods, such as *Acanthostega*, these bracing struts – which were called the hyomandibulars – were sturdy girders. They didn't conduct sound all that well, especially anything above a basso roar, as if of distant thunder.[1]

When the tetrapods eventually made landfall, they found themselves in the completely different acoustic environment of the open air. The cartilages that formed the gill arches became supports to the tongue and the larynx. Only the hyomandibulars remained in place. But now they became dedicated to the sensing of sound. The spiracles were roofed over by thin membranes. These were the eardrums. The hyomandibulars conducted vibrations from the eardrums direct to the inner ear. Because of this new role, the hyomandibular acquired a grander name: the *columella auris* – the small ear column. But it is also known, less grandly, as the stapes, or stirrup bone. The stapes lay between the eardrum at one end, and the inner ear at the other. Its pocket-sized empire was the middle ear.[2]

As sounds beat on the eardrum, vibrations are conducted through the stapes to the inner ear. This is how amphibians, reptiles and birds hear to this day. Over time, the stapes became became whip-thin and sensitive to a whisper. Even so, it had its limits. The tweets, croaks and burbles of birds fill the air – birds make some of the loudest noises heard in nature.[3] Yet birds are largely insensitive to sounds of a frequency higher than about 10,000 cycles per second, or 10 kilohertz (kHz).[4]

Mammals, however, do it differently. Rather than having just one bone in the middle ear – the stapes – they have three. The stapes connects to the inner ear as before, and thence to the brain, but two other bones have squeezed in between the eardrum and the stapes. These are the malleus ('hammer'), which is attached to the inside of the eardrum; and the incus ('anvil'), that links the malleus to the stapes.[5]

The effect on mammalian sensation has been dramatic. The chain of three bones acts to amplify sound. It also increases the sensitivity of the ears to higher frequencies. We humans, at least in childhood, can hear notes as high as 20 kHz, much higher than the highest song of the skylark.[6] But humans are cloth-eared compared with many other mammals, such as dogs (45 kHz[7]), ring-tailed lemurs (58 kHz[8]), mice (70 kHz[9]) and cats (85 kHz[10]); and profoundly deaf compared with dolphins (160 kHz[11]). The evolution of the chain of three bones in the mammalian middle ear opened up to mammals an entirely new sensory universe, inaccessible to other vertebrates.

It was as if they had stumbled across a small hole in the high hedge surrounding the dense woodland to which they had been accustomed, only to discover open fields of a spaciousness they had never imagined possible.

Where did the malleus and the incus come from?

When jaws first evolved, in the fish fleeing for their lives from other denizens of the deep, the jaw joint came to rest just underneath the spiracle, the remnant gill slit that would, in tetrapods, become the eardrum. It was one of those quirks of time and chance, then, that the jaw hinge just happened to be close to the ear, rather than anywhere else.

The jaw joint and the eardrum are, however, more than just near neighbours. They are intimate with each other. This intimacy was to be a key to the eventual success of mammals.

When the lower jaw first evolved, it was a rod of cartilage, one half of the first gill slit that had bent double to form the jaws. The upper half became the upper jaw, the lower half, the lower. Over time, this cartilage turned into bone: although a remnant – the Meckel's cartilage – persists, at least in the embryo, as a thin strap of tissue on the inner surface of the lower jaw, before it fades away.

The lower jaw of a reptile, or a dinosaur, is a complicated thing. It is made of not one bone, but several, each of which has its own task. The dentary is the bone near the front which, as the name suggests, carries the teeth. The articular, in contrast, is near the back, and forms the jaw hinge – or articulation – with a bone in the base of the skull called the quadrate. The same was true of the therapsid ancestors of mammals.

As the therapsids evolved into mammals, becoming ever smaller, from animals the size of large dogs, to small dogs, to cats, to weasels, to mice, to even smaller shrews, and all the time becoming furrier, the jaw changed, too. The dentary started to assume a greater role in the jaw as a whole. Like the outsize cuckoo chick that forces its unwitting step-siblings out of the nest, the dentary pushed backwards, so that the other bones in the jaw were either completely absorbed by it, or kettled into a small enclave at the back, next to the stapes. The dentary moved so far back, in fact, that it formed its own entirely separate hinge with the skull, with a different skull bone, the squamosal.

The effect of this was to relieve the quadrate of its duty as a hinge joint. Being close to the stapes, it became recruited instead as an ear bone. It became the incus. The articular bone was the next to follow, becoming the malleus.[12]

In some of the precursors of mammals, the jaw joint was an uneasy composite of dentary and squamosal, and quadrate and articular. Because the quadrate and articular were evolving into the incus and malleus, they had to do two entirely separate jobs. In one, they had to form part of the jaw suspension, which demands robust strength. In the other, they had to conduct sound, which requires sensitivity. As with the stapes, many millions of years earlier in the fishy ancestors of tetrapods, this was a compromise that could not hold.

Eventually, the quadrate and articular were left to float freely in the middle ear – at first, attached to the jaw by a wisp of the retreating Meckelian cartilage. Then even that disappeared. With the evolution of the mammalian middle ear, mammals awoke to an acute sensitivity to the world of sound no tetrapod had ever experienced.

The mammalian middle ear evolved as a direct consequence of the drive to small size[13] – and it evolved not just once, but at least three times, independently; in the ancestors of the animals that would become the egg-laying platypus and echidna of Australasia; in the ancestors of the marsupials and placental mammals, which together make up more than 99 per cent of all mammal species alive today; and again in the multituberculates, a group of mammals that looked rather like rodents and lived from the Jurassic to the Eocene, when they became extinct.

The long journey from therapsid to mammal started in the earliest Triassic, with cynodonts such as *Thrinaxodon*. This creature was a Jack Russell terrier seen through half-closed eyes. However, apart from its short, stumpy tail and rather sprawling waddle, it was amazingly mammal-like. It had whiskers, and fur.[14] It was a digger of holes and burrows.

The differences were even more marked inside. Even at this early stage, the dentary dominated the lower jaw, although the middle ear still consisted of the stapes, alone.

In reptiles, the teeth are simple points, and are replaced whenever one falls out. Pelycosaurs had shown a penchant for varying the shapes and sizes of their teeth, creating a whole canteen of cutlery, each utensil specialized for a different task. This trend continued with their therapsid descendants.

One thinks of the gorgonopsians, with their oversized canines; the dicynodonts, too, their prey, with their effective combination of canine teeth and horny beak. Cynodonts had canines too – their name means 'dog tooth' – but they continued the trend towards differentiation in the other teeth. Mammals have four basic kinds of teeth: nippers (incisors), stabbers (canines), slicers (premolars) and, at the back, crushers (molars). *Thrinaxodon* had nippers and stabbers, but the teeth behind the canines were not clearly differentiated.

Rather than bearing ribs all the way along the vertebral column, as reptiles do, the ribs of *Thrinaxodon* were confined to the thorax, what we would today call the ribcage. This is a uniquely mammalian feature. It implies that *Thrinaxodon* had a diaphragm, a sheet of muscle that divides the thorax from the viscera internally, and which would have allowed for much more powerful and regular breathing.[15]

Another adaptation for breathing was inside the nose. In contrast to reptiles, in which the internal nostrils empty into the roof of the mouth cavity, near the front, *Thrinaxodon* had developed a long nasal cavity almost entirely separate from the mouth cavity, meeting it only at the very back, so air could pass cleanly to the throat and avoid all the mastication in between. This meant that the animal could chew its food without having to pause for breath. The enlarged nasal cavity became filled with a labyrinthine scrollwork of bones, supporting a large area of mucous membrane – implying a keen sense of smell, and the ability to warm air breathed in while chomping on some benighted animal even smaller than itself.

The picture that emerges is of an active animal with a fast-running metabolism: parallel to that of dinosaurs, but achieved in a different way. Rather than a full-body network of air sacs, the diaphragm pumped air in and out. Like the smaller dinosaurs, *Thrinaxodon* and later cynodonts conserved heat with a fur coat. Because a fast metabolism takes a lot of fuel, the act of eating became more efficient. Rather than swallowing food whole and digesting at leisure or, as in birds and dinosaurs, by grinding it up with stones in a crop, or gizzard, *Thrinaxodon* used a battery of different teeth to slice and dice its prey while it was still in the mouth, exploiting its ability to breathe while chewing.

The transformation between cynodonts and the earliest mammals was an ongoing affair that involved several different and diverse lineages of therapsids. By the late Triassic, animals had appeared that were indistinguishable from mammals in all important respects. And they were tiny: *Kuehneotherium* and *Morganucodon* were no larger than modern shrews, perhaps 10 centimetres long at most. They had fully formed middle ears,[16] and their teeth had evolved, too, into clear nippers, stabbers, slicers and crushing molars.

The molars were special, in that rather than having all their pointed cusps in line, as in a shark tooth, the cusps tended to be staggered, to create a two-dimensional chewing surface, with the various cusps and pits in the lower molars interlocking with those in the upper jaw. This made for even more efficient food processing, yet another weapon in the armoury for small, hot-rodded creatures that had to eat a large fraction of their own body weight in insects each day, simply to stay alive. Even at this early date, though, each mammal had its own dietary speciality. Whereas *Morganucodon* could tackle hard and crunchy prey, such as beetles, *Kuehneotherium*'s tastes ran to softer items, such as moths.[17]

The fast metabolism, fuelled by efficient chewing and breathing, part of which, incidentally, resulted in an improved sense of smell; the seemingly unswerving tendency towards ever-smaller body size, which in turn prompted the evolution of acute high-frequency hearing; and the habit of hiding away in burrows. All speak to mammals evolving into a habitat that had been closed to almost all other vertebrates – the night.

Pangaea in the Triassic was in many ways a hostile place. Away from the storm-wracked shores of the Tethys Ocean, much of the land was desert in which, by day, the ground was too hot to touch. *Kuehneotherium* and *Morganucodon* lived in such a desert, between 20 and 30 degrees north of the Equator. In this environment, the best strategy is to hole up in a burrow well below the surface, out of the day's heat, and emerge to hunt at night, or very early in the morning. To do this, a fast-running metabolism is essential. Reptiles that rely on the heat of the sun to warm up will be beaten to the juiciest insects by mammals that are already warm. The insects, too, will tend to be more torpid at such times, making them easier prey.

For animals that spend their days in dark burrows, emerging only at night to hunt beneath the stars, vision is far less important than hearing, touch and smell – senses that had been slowly improved in the therapsids since the days of *Thrinaxodon*. In the mammals, they reached their acme. Above the ground, and by day, the Triassic was a riot of reptiles. But the night belonged to the mammals. It was to be their playground for the next 150 million years.

Every dinosaur that ever lived hatched from an egg. The same was once true of mammals. It was a good habit to have, for, as we have seen, eggs allow for very rapid production of numerous offspring with rather little parental investment. *Kayentatherium*, a very mammal-like therapsid that lived in the Jurassic – and therefore one of the very last therapsids that wasn't a fully furred mammal – hatched dozens of young, each of which looked like a miniature adult, ready to make its own way in the world.[18]

But a change was going to come. That change was in the brain. For early mammals were evolving larger brains. Hatchlings began to look like what we imagine baby animals look like – relatively undeveloped, with large heads relative to their bodies, full of burgeoning brains. Brain tissue is very expensive to make and maintain, and puts an enormous strain on small animals already running as fast as possible just to stay in the same place. So, rather than laying lots of eggs, mammals had smaller clutches, and devoted more time to caring for each hatchling. Females began to secrete a fat- and protein-rich substance from modified sweat glands, ensuring that the young were fed a diet packed with all the nutrients they would need for rapid growth. We call this substance 'milk'. Historically and etymologically, the presence of milk-producing organs or *mammae* is what makes a mammal a mammal.

The life of a mammal was one of tension. By the time the dinosaurs appeared, late in the Triassic, mammals were well on the way towards refining the art of being small, and living short, high-pitched, action-packed lives. But the energetic strain would have been made easier had they been they able to return to something like normal size, especially now that they had large brains to support.

The problem was that by the time the mammals were in a position to expand out of the role of the small, nocturnal insectivore and scavenger, the dinosaurs had evolved to fill all the available ecological slots. Indeed, to a small, intelligent, active dinosaur, mammals were more than just competition – they were prey.

Not that the mammals didn't make several attempts to break out. Animals that live fast also evolve fast. During the time of the dinosaurs, at least twenty-five different groups of mammals evolved.

They were a venturesome bunch, and not to be kept down. Although mammals during the reign of the dinosaurs were never very large, some did evolve to the size of opossums, even badgers – large enough to steal dinosaur eggs and baby dinosaurs,[19] and perhaps prompting some of the smaller, feathered dinosaurs to remain up in the trees.

If they did, they would have shared the habitat with not one but at least two entirely separate kinds of mammal that evolved into something like a flying squirrel.[20] Neither was the water safe: the 800-gram *Castorocauda* had a flattened, beaver-like tail, a furry pelt and sharp teeth, perfect for diving for fish in Jurassic ponds.[21] Madagascar, ever a haven for the unusual, hosted the rabbity *Vintana* and *Adalatherium*, which, with large eyes and a keen sense of smell, would have been alert to the merest twitch of a predatory dinosaur.[22]

By the time dinosaurs died out, all but four of these lineages, bubbling and vivacious as they were, had become extinct. The survivors were the egg-laying monotremes, the marsupials, the placental mammals, and the multituberculates. Each could trace roots into the rich mould of an evolutionary story that already ran deep.

Monotremes are mammals in that they suckle their young, even though they hatch from eggs. This group, represented today by the platypus and echidna of Australasia, is the last and peculiar remnant of a very ancient lineage of mammals that went its own way in the Jurassic period, and was found all across the southern continents.[23]

Most other mammals – the placental mammals – gave up the egg-laying habit entirely, nurturing a smaller number of young internally. Mammal embryos have all the same membranes as the amniotic egg – just without the shell. In the ultimate act of selfless devotion, the mother herself has taken on that protective role. Like that of the monotremes, the lineage of the placental mammals can be traced back a long way, in this case to small, tree-climbing creatures that hunted insects in the branches of Jurassic forests.[24]

The marsupials, though, evolved a canny compromise, between the set-and-forget egg laying of monotremes, and the full-on internal nurturing of placental mammals. Although marsupials nurture their young internally, the offspring is birthed when it is hardly more than an embryo. Once in the outside world, the tiny creature crawls through the forest of its mother's fur into a pouch and attaches itself to a teat. There, secure and suckling, it develops to term. This strategy is an adaptation to harsh, marginal environments where food is hard to come by. If beset by troubles, a pregnant marsupial can abort its offspring and create new ones later, if and when circumstances allow.

Marsupials are as antique as placentals in the fossil record,[25] and have a long and illustrious history. They have done especially well when confined to island continents, where they have assumed an astonishing range of forms. For much of the Cenozoic era they had South America as their own fiefdom, which they shared with the very strange (and placental) edentates – sloths, anteaters, armadillos and allies – but which was ruled by animals such as *Thylacosmilus*, a marsupial version of a sabre-toothed tiger, and its outriders, the borhyaenids, which varied in size and manner from wolves to bears. When South America collided with North America, an invasion of placental mammals from the north all but wiped them out.

A few South American mammals fought back, however, with counter-invasions, spearheaded by giant ground sloths and armadillos, with a vanguard of opossums, which continue to raid American trash cans to this day. Most marsupials today live in Australia, where their unique mode of reproduction suits them to the parched interior of that increasingly arid continent.

So, by the time the dinosaurs finally died out, the mammals were ready, honed by a million years of evolution. They burst forth like a well-aged champagne, shaken beforehand, and inexpertly uncorked.

Waiting for them, though, were the top predators of the immediate post-apocalypse world. These were birds, the phorusrhachids. Immense, flightless relatives of cranes and rails, with horsehead-sized skulls, they decapitated any mammals foolhardy enough to leave their burrows. It was as if *Tyrannosaurus rex* had returned.

Even these horrors disappeared into the dust of the Palaeocene plains, and mammals – placental mammals especially – expanded, in size and form. The first waves, though, seemed shambling and unformed, as though undecided as to their purpose. Animals such as pantodonts and dinocerates, arctocyonids and mesonychians, all now long vanished, combined features of carnivore and herbivore. Pantodonts and dinocerates were herbivores, and among the earliest to grow to large sizes. Some pantodonts were as large as rhinos;

some dinocerates were as big as elephants. Although plainly herbivores, they had fearsome canine teeth.[26] The arctocyonids combined the canine teeth of bears with the hooves of deer.

Equally ambiguous were the mesonychians. The giant terror birds (the phorusrhachids) met their match in the mesonychian *Andrewsarchus*, a terrifying animal, man-high at the shoulder, with a head as wide as an Alaskan brown bear's is long, and which could have snorted the entire skull of a wolf up one nostril. And it had hooves on its feet. *Andrewsarchus* looked like a very large, very angry pig.[27]

The Earth at the end of the Cretaceous was a warm and balmy place, despite the asteroid; and the warmth continued. But as the Palaeocene wore on and became the Eocene, equable warmth became torrid heat. Plains and woodlands became jungles. The first wave of early and ambiguous mammals was gradually replaced by others that had clearer sets of life goals. Ungulates – the hoofed mammals – made their first appearance, but in those days, the Eocene, they were small, and looked more like squirrels, scampering and scurrying among the towering trees, possibly to avoid predators such as *Titanoboa*, a snake the size of a bus.[28]

Some of the earliest even-toed ungulates escaped in the most unlikely direction imaginable – they returned to the water, and became whales. They did this, moreover, with enthusiasm, and, in evolutionary terms, great haste.

The first hints of whaleness – in the running, predatory,

wolf-like *Pakicetus* and fox-sized *Ichthyolestes* – were seen in the rather long, toothy jaws, a feature often found in fish-eaters; and various crinkles in the anatomy of the inner ear that might predispose towards hearing in water.[29] More obviously aquatic was the more sea lion- or otter-like *Ambulocetus*, with shorter (though still fully functional) limbs.[30]

But it wasn't long before whales had become fully aquatic, in forms such as the 20-metre-long *Basilosaurus*, every inch the coiled sea serpent of lore, even though it retained tiny vestiges of its hindlimbs as a memory of its terrestrial forebears.[31]

After that, there was no stopping them. Whales replaced the giant sea lizards – a niche that had been vacant since the extinction of the plesiosaurs and mosasaurs at the end of the Cretaceous. They became highly successful mammals, numbering among the most intelligent of all animals, as well as, in the blue whale, the largest animal that evolution has ever produced. But perhaps more remarkable than their transformation itself was its speed – from fully terrestrial doglike runner to fully marine in just 8 million years.[32]

Another transformation was perhaps even more startling, in that it seems to have erased almost all traces of its ever happening.

The separation of Africa from South America during the Cretaceous period left Africa an island continent. It remained cut off for around 40 million years. The early, insectivore-like placental mammals of Africa, left to themselves, diversified to such an extent that all external signs of their common heritage disappeared.[33] And diversify they did, into magnificent elephants; the aquatic sirenians such as the dugong and manatee; the aardvark, tenrecs, the golden mole, elephant shrews and hyraxes. All are Afrotheria, a parallel radiation to a more northern group, the Laurasiatheria – which includes ungulates, whales, carnivores, bats, pangolins and the remaining insectivores.

In every classification, there's always a group of leftovers. In the case of the mammals, these were the Euarchontoglires – an assortment of scrappers that included rats, mice, rabbits, and, seemingly almost as an afterthought, the primates. These small, scampering creatures, with forward-facing eyes, colour vision, brains inclined to curiosity and hands to exploration, peered out from the towering tropical forests of the Eocene at a fast-changing world.

Timeline 4. The Age of Mammals

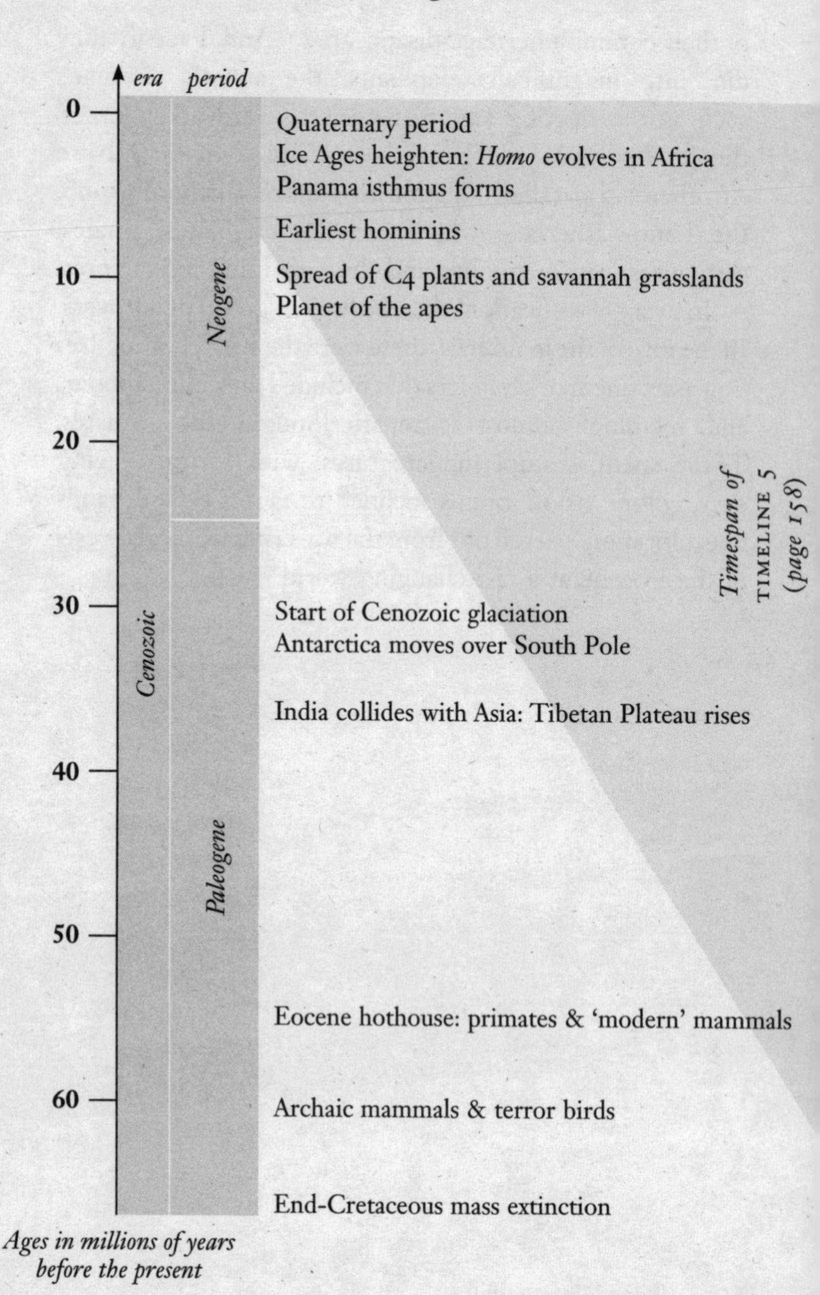
era
period
0
10
20
30
40
50
60
Neogene
Paleogene
Cenozoic
Quaternary period
Ice Ages heighten: *Homo* evolves in Africa
Panama isthmus forms
Earliest hominins
Spread of C4 plants and savannah grasslands
Planet of the apes
Start of Cenozoic glaciation
Antarctica moves over South Pole
India collides with Asia: Tibetan Plateau rises
Eocene hothouse: primates & 'modern' mammals
Archaic mammals & terror birds
End-Cretaceous mass extinction
Timespan of TIMELINE 5 *(page 158)*
Ages in millions of years before the present

9

PLANET *of* *the* APES

The choreography of continental drift is as remorseless as it is slow.

By around 30 million years ago, the continent of Antarctica had broken free from Pangaea and had drifted so far south that it was completely surrounded by ocean. The effect of this single event on the Earth's climate was profound and long-lasting. For the first time, it was possible for an ocean current to swirl uninterrupted around the new continent. This current prevented water warmed in the tropics from reaching Antarctica's then-mild shores. A chill hung over the jagged, tree-mantled teeth of the Transantarctic Mountains, one of the most formidable ranges on the planet.

It came to pass that one year, snow that had fallen during the winter failed to melt completely by the following spring, but lay on the ground all year long. More snow piled up, snow on snow, century upon century, crushed into unmelting, unremitting ice. Glaciers began to form in the high valleys.

As Antarctica continued to drift southwards, the sun stood ever lower at midsummer, and the nights of winter became longer. Eventually, the year came when the winter sun failed to rise at all, and the continent spent six months in unbroken darkness. The glaciers grew so great that they buried and overtopped the ranges of mountains in whose valleys they had been nurtured. Walls of ice ploughed into the lowlands, erasing everything in their path. The coastline was no barrier. The ice marched on, into the ocean, forming shelves of ice over the sea, calving icebergs to chill the surrounding ocean still further.

Within a few million years, a continent that had once been lush and green had become a dry, icy desert too harsh for all but the most primordial kind of life – lichens, mosses – and even then, only in the most sheltered, north-facing spots on the edges of the landmass. The seas surrounding it, though, swarmed with life.

It was a similar tale in the far north, if curiously reversed. The northern continents, drifting northward still, surrounded the Arctic Ocean so that very little warm water could reach it from further south. A permanent ice cap began to form in the northern sea, as if aping the much greater one on land, in the far south. After many millions of years of being completely free from polar ice, permanent ice caps had returned to the Earth.

The consequences were felt worldwide. Whereas once the world had been tolerably warm almost everywhere, a steep gradient in climate grew between the poles and the tropics. Winds blew up. Climate became more variable, more seasonal, and cooler.

It spelled the end for the jungle planet that the first primates called home.[1]

The jungles broke up into isolated fragments of woodland. Great plains started to appear in the spaces between, clothed in a new kind of plant – grass.[2] Growing from the bottom up, rather than the top down, grass could be cropped continually without being killed. This strange new gift was soon exploited by animals that evolved to graze it. But grazing is harder work than nipping the tender leaves of jungle trees, which the animals had done before. Grasses are rich in silica, a mineral that literally sands down an animal's teeth as it chews.

The ungulates had evolved as forest browsers, but now grew deeper jaws and many-cusped teeth capable of cropping this gritty fare. As they evolved, they grew, and there was thunder on the plains under the hooves of horses and the feet of gigantic rhinos.

The descendants of small, hippo-like creatures that had browsed the swamps and wetlands of Africa moved onto the dry, hard ground and became elephants. Larger and ever more mighty as the ages rolled on, they came to the savannah. The herds brought predators in their wake.

The primates, too, adapted. Although many remained in the shrinking forests, living an ever more marginal existence, some began to supplement life in the trees with episodes on the ground. Like the ungulates, they too became larger: the scampering monkeys became a more considered deliberation of apes.

By the Miocene, the Old World had become a Planet of the Apes. The thinning forest patches and the surrounding drylands rang with their hoots and calls. *Ouranopithecus*[3] swayed in Greece, while *Ankarapithecus*[4] swung in Turkey. *Dryopithecus* patrolled central Europe; *Proconsul*, *Kenyapithecus* and *Chororapithecus* browsed Africa, where a relative of the last evolved into the gorilla.[5] There was *Lufengpithecus* in the forests of China and *Sivapithecus* in southern Asia, where its relatives eventually retreated to the last jungles and, via Thailand's *Khoratpithecus*,[6] became orangutans.

Some of these apes were so large that they could no longer run along the branches that had once been their highways.[7] Instead, they adopted a variety of different postures, such as hanging beneath branches from long arms, or a mixture of climbing and clambering. Over time, some such as *Danuvius*, from central Europe, came to adopt a more upright stance.[8]

Not all these essays were entirely successful over the long term. *Oreopithecus*, marooned on an island in the Mediterranean that would one day become Tuscany, experimented with walking upright.[9] It died out, nonetheless.

And still the Earth cooled. The forests shrank even further, driving most of the remaining apes into jungle refuges in the deep forests of Central Africa and South East Asia.[10] For the rest, the choice was stark: final banishment from Eden, or extinction. The refugees took little with them but a tendency to get up on their hindlimbs and walk.

By 7 million years ago, the descendants of Eden had become better walkers than climbers. The cooling climate had transformed monkeys into apes; and apes into something else again. As so often before, the restless Earth, turning in slumber, rucked up its thin coverlet of life, and life did its best to hang on. The remaining apes, driven by forces mightier than any of them could possibly have imagined, took their first steps on the long journey towards humanity.

Walking upright as of habit rather than merely on occasion is the earliest mark of the hominins – the human lineage.[11] The earliest hominins emerged in the late Miocene, around 7 million years ago. One such was *Sahelanthropus tchadensis*,[12] a creature that foraged along the shores of Lake Chad in West Africa. The region was once lush: the lake one of the most extensive in the world. But the tendency for the climate to become dryer has only increased in the meantime: the lake has shrunk to a tiny vestige of its former self, and the surrounding countryside has become an inhospitable and blasted desert.[13] *Sahelanthropus* was not alone. In East Africa, some 5 million years ago, lived other bipeds such as *Ardipithecus kadabba*[14] from Ethiopia, and *Orrorin tugenensis*, from Kenya.[15] For the primates, walking upright, like most other innovations in human prehistory, began in Africa.[16]

Standing and walking is, for us, so easy, so natural, that we take it for granted. Many mammals can stand upright for a short time, and even walk. But it takes effort, and they soon slump back to all fours, the typical mammalian state.[17] Hominins are different. Walking upright is their default – locomotion on all fours, using hands and feet to walk, is, in contrast, unnatural and difficult. The adoption of bipedality, by a lineage of apes living in the river margins and in the wood-eaves of Africa 7 million years ago, was one of the more remarkable, unlikely and puzzling events in the entire history of life. It required a complete re-engineering of the entire body, from head to toe.

In the head, the hole where the spinal cord enters the skull shifted from the back (where it is found in quadrupeds) to the base. This feature, if not much else, marked *Sahelanthropus* as a hominin. It meant that its face was directed forwards, rather than upwards at the sky, when it walked on its hind limbs, and that the skull balanced atop the spinal column rather than being cantilevered from one end.

The effects on the rest of the body were likewise profound. When the backbone evolved half a billion years ago, it was a structure held horizontally, in tension. In hominins, it moved through ninety degrees, to be held vertically, in compression. No more radical alteration in the engineering requirements of the backbone has happened since it first evolved, and it can only be regarded as maladaptive – witness that back problems constitute one of the most costly and frequent causes of illness in humans today. Dinosaurs made a huge success of being bipeds, but did so in a different way – they held their backbones horizontally, using their long, stiff tails as counterbalances. But hominins, like apes, have no tails, and achieved bipedality the hard way.

Matters were worse for pregnant females, which had to adjust to an increasingly unstable and ever-changing load – a circumstance that has left its imprint in human evolution. And no wonder, given that for most of human history, adult females – on whom the continuation of the species depends – spent much of their lives either pregnant or nursing.[18] And worse still: hominin legs tend to be longer, as a proportion of overall height, than they are in apes. Longer legs make locomotion more energy-efficient, but there is a cost. The foetus is even higher off the ground than it might be, raising the overall centre of mass, and increasing instability.

If that weren't enough, a hominin has to move by lifting one foot from the ground, shifting the centre of mass sharply, and then correct this before it falls over – and it has to do this with each step it takes. This requires a quite remarkable degree of control, in which the brain, nerves and muscles work flawlessly together to such an extent that we aren't aware of it.

The first hominins seemed puny compared with some of the animals with which they shared their world. In fact, they were the elite fighter jets of the animal kingdom. Quadrupeds can rumble along, run fast, and even turn rapidly, but such acts often require torque driven by a long, swishing tail, as in the cheetah as it hunts.[19] In general, animals with a leg at each corner are like work-horse cargo planes, which, when pointed in the right direction, keep gamely flying along. Human beings, without such aids, are like fighter jets – almost preter-naturally manoeuvrable, at the cost of stability: only the very best pilots get to fly the fastest jets. Hominins not only walked, like the dinosaurs; they danced, they strutted, they pivoted and they pirouetted.

The gains achieved by bipedality were eventually enormous. But the wonder is how it got started at all. It is testament to the unlikelihood of bipedality as a proposition that hominins are among the very few mammals that go on two legs as a normal part of life[20] – a rarity piqued by the helplessness of any human suddenly deprived of the use of one of its hindlimbs.[21] Once hominins started on the little-frequented path that led towards bipedality, natural selection ensured that they had to become very good at it, very quickly.

Human walking is one of the great underappreciated wonders of the modern world. Today, scientists are capable of teasing out the structure of subatomic particles, detecting the rumble and squeak of merging black holes millions of light years away, and even peering into the very beginnings of the universe. Yet no robot has been created that can mimic the natural grace and athleticism of an ordinary human being as it walks.

The question remains – *why?* The easy answer is that bipedality is just one of many peculiar modes of locomotion that apes have tried over millions of years, including swinging using elongated arms, as in gibbons; clambering using all four limbs as hands, as orangs do; and the unique quadrupedal knuckle-walking of chimps and gorillas. But why hominins tried bipedal walking, rather than any other mode of getting from one place to another, remains an open question. Certainly, life in open country does not require it. Many large monkeys, such as macaques and baboons, live in open country and remain with all four feet firmly planted on the hard, dry ground.

Suggestions that bipedality freed the hands to, say, make tools, or hold infants, don't wash either, given that many animals manage both without the thoroughgoing change to bipedality in which hominins are invested. As far as the earliest hominins are concerned, the most that can be said is that they might have been somewhat pre-adapted to it by virtue of a kind of upright climbing-and-clambering mode of getting around they had begun to adopt in trees, after which walking on the ground would not have been such a great change. To them, walking may have been like clambering around in branches – but without the branches.

In any case, many still retained the capacity to climb. The feet of one of the earliest, *Ardipithecus ramidus*, which lived 4.4 million years ago in Ethiopia,[22] had divergent big toes, which, like thumbs, suggest a capacity for gripping – the mark of a creature more at home in the trees than walking comfortably in their shade.[23] Another species, *Australopithecus anamensis*, which lived in East Africa between 4.2 and 3.8 million years ago, was likewise primitive in many ways, but more sure-footed on the ground.[24]

Australopithecus anamensis overlapped in time with a range of other, similar species. One of these, *Australopithecus afarensis*, lived in the same region between 4 and 3 million years ago,[25] and was a better biped still. This was one of the most successful of all these early hominins, in that it ranged beyond East Africa and was found as far west as Chad.[26] Wherever it roamed, it did so as upright as we do now,[27] though it was still a capable climber.[28]

None of this should give the impression that a series of ever more bipedal species replaced one another in some orderly, preordained fashion. The hominins were spread thinly over the savannahs of East Africa, preferring to live in a mixed environment of grassland, woody scrubland and shady woodland close to water,[29] with some species happier in trees than others. As late as 3.4 million years ago, tree-hanging hominins similar to *Ardipithecus* still hung around in the woods.[30]

For all these early hominins, then, walking upright was a part of a daily routine that included climbing, and, perhaps, nesting in the branches of trees, as apes do today. The mixture extended beyond environment to diet. Some hominins were beginning to incorporate harder food items, such as nuts and tubers, into the traditional primate diet of fruit, young leaves and insects. The evolutionary response led to changes comparable to those seen in the savannah ungulates: flaring cheekbones to accommodate huge chewing muscles, deep jaws and teeth like tombstone slabs. Several species of this highly specialized type, loosely grouped in the genus *Paranthropus*, appeared in Africa between about 2.6 million and 600,000 years ago. These quintessentially savannah creatures lived alongside more generalized hominins – various species of *Australopithecus* and our own genus, *Homo*[31] – some of which had become fond of more succulent fare.

At some time around three and a half million years ago, some of these early hominins developed a taste for meat, usually scavenged from kills made by other animals. No early hominin had the teeth or claws to match a lion or leopard – but they had begun to chip at rocks, make sharp-edged tools, and were developing the art of butchery.[32]

The earliest tools were no more than chipped rocks.[33]

But their impact on human life was to be profound. Acute primate binocular vision inherited from Eocene tree-living ancestors – combined with a rock thrown with hands freed from mundane locomotion – could brain a scavenging lion, or scatter vultures from a carcass. Even before the development of cooking, the use of these same simple stone tools to slice meat and pound vegetable matter significantly increased the nutrients available to creatures[34] that had to use endless ingenuity to keep the perpetual threat of starvation at bay. Meat, and marrow liberated from stone-shattered long bones, are packed with vital proteins and fats, and could be digested more easily than fibrous roots and nuts, which called for relentless chewing. The hominins that ate meat and fat evolved smaller teeth, and smaller chewing muscles. The energy they saved went into growing larger brains; the time, into doing things other than collecting food and chewing it.

Hunger, though, was never far behind. It occurred to some of these hominins in their moments of leisure that meat might be more succulent were it to be caught fresh, rather than their having to rely on scraps already much chewed over by other animals. They learned to make better stone tools.

Most of all, they took a step that would be as revolutionary as standing upright had been to their now-distant forest ancestors: they learned how to run.

Timeline 5. Humans Emerge

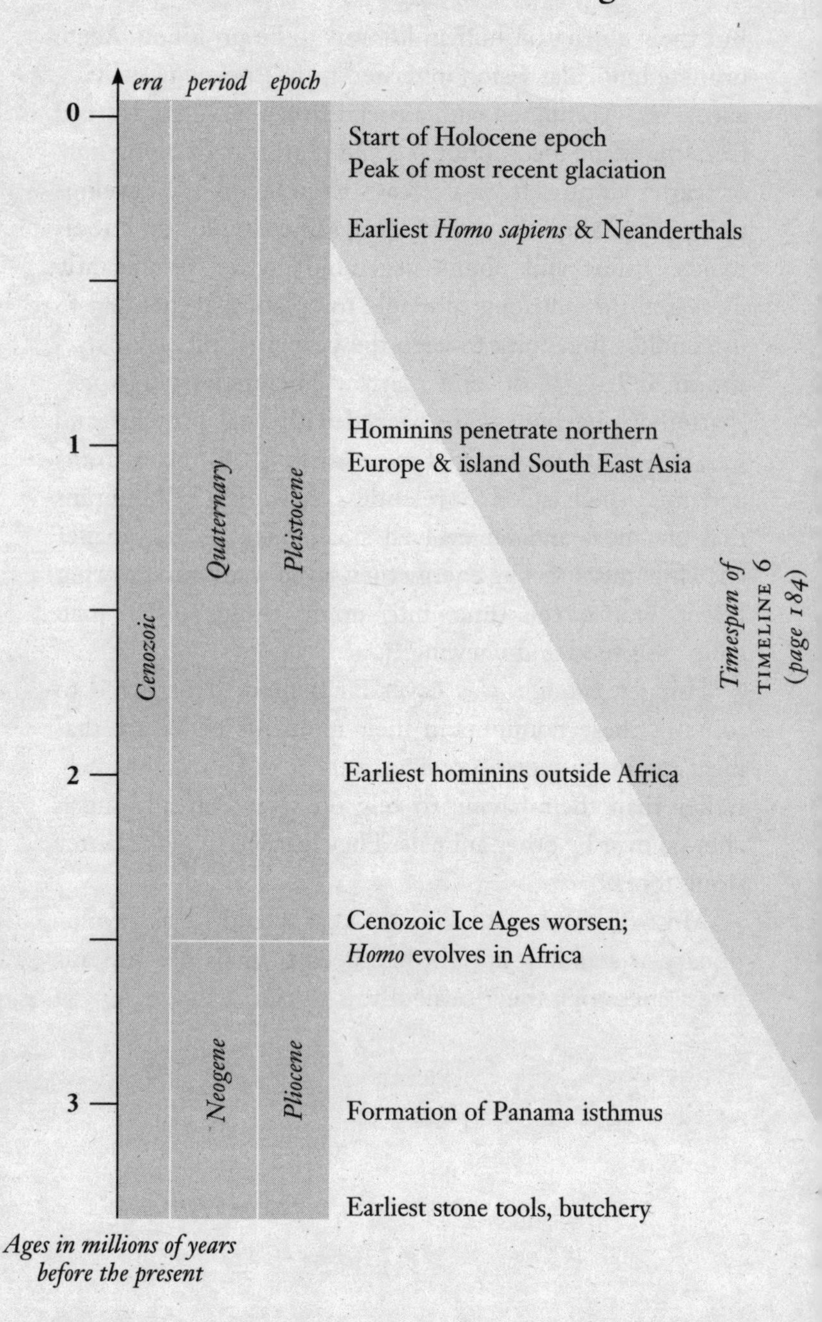

era
period
epoch
0
1
2
3
Cenozoic
Quaternary
Neogene
Pleistocene
Pliocene
Start of Holocene epoch
Peak of most recent glaciation
Earliest *Homo sapiens* & Neanderthals
Hominins penetrate northern Europe & island South East Asia
Earliest hominins outside Africa
Cenozoic Ice Ages worsen; *Homo* evolves in Africa
Formation of Panama isthmus
Earliest stone tools, butchery
Timespan of TIMELINE 6 *(page 184)*
Ages in millions of years before the present

10

ACROSS *the* WORLD

After more than 50 million years, the long, slow decline in the Earth's climate was about to reach its nadir.

Everything was in place.

In the far south, a circumpolar current had locked Antarctica in the deep freeze. In the far north, the converging continents had imprisoned the Arctic Ocean in its own kind of frigid hell. But more was to come.

The knell came from space. Not as a sudden impact, like the one that had brought the reign of the dinosaurs to such a fiery end – but as a series of almost imperceptibly tiny changes in the way the Earth orbited the Sun. Such changes had always been there, in the background, but their effects on the planet's residents had almost always been too small to matter. All that was about to change.

The Earth's orbit around the Sun is not perfectly circular, but very slightly elliptical. If it were circular, the Earth would always remain a fixed distance from the Sun. But because the orbit is elliptical, the Earth's distance from the Sun varies during the year: sometimes it is closer to the Sun, sometimes further away. This deviation from perfect circularity is known as eccentricity, and it is driven by the

Earth's gravitational interaction with the other planets as they make their own journeys about the Sun.

At its closest approach, the Earth is 147 million kilometres from the Sun. At its most distant, 152 million kilometres. This counts for very little in the great scheme of things. However, at times, the Earth's orbit becomes more eccentric – stretched – so that our planet will come as close as 129 million kilometres to the Sun, and range as far away as 187 million. It is as if the Earth's orbit slowly 'breathes' in and out. Each complete breath lasts 100,000 years. The more stretched the orbit, the more extreme the climate, as the Earth gets much nearer to the fires of the Sun than it otherwise might; and ventures further away, into the long dark of deep space.

At the same time, the tilt of the Earth's axis wobbles with respect to its plane of revolution about the Sun.

Seasonal change, and the division of the Earth into climatic bands, are all consequences of the axial tilt of the Earth. During the northern summer, the North Pole is tilted towards the Sun, at an angle of 23.5 degrees from the vertical. This means that everywhere north of latitude 66.5 degrees[1] – that is, the Arctic Circle – is continually bathed in sunshine. By the same token, in the northern winter, when the northern hemisphere is tilted away from the Sun, the Arctic languishes in total darkness. In the southern hemisphere, and the Antarctic Circle, at 66.5 degrees south, it is the other way round. The Tropics of Cancer and Capricorn, at latitudes 23.5 degrees north and south respectively, mark the furthest north or south from the Equator that the Sun stands directly overhead at midday.

The current value of 23.5 degrees is something of a happy medium. Axial tilt can vary between 21.8 and 24.4 degrees, with a period of about 41,000 years. The degree of tilt affects seasonality. When the tilt is greater, summers will be – on average – slightly hotter, the winters colder; the realm of the Arctic and Antarctic will be greater; and, in the Tropics, the Sun will stand overhead at noon in midsummer at a higher latitude. In other words, the Earth's climate becomes, very slightly, more extreme. When the tilt of the axis is less than 23.5 degrees, the climate is generally milder.

A third cycle is precession, in which the tilted Earth's polar axis itself revolves, though much more slowly than the daily rotational cycle – much like the way the axis of a spinning top revolves as the top spins. This cycle takes about 26,000 years to complete. It can be seen, for those sufficiently patient, by a slow movement of the pole that describes a circle around the sky. At the moment, the North Pole seems to point, more or less, to the star Polaris, the 'Pole Star', in the constellation of Ursa Minor. Because of precession, however, Polaris will in time be replaced by Vega, in the constellation Lyra, another prominent northern star.[2] This will be clearly visible to anyone content to wait 13,000 years.

As a consequence of these three cycles, each supplementing the others, the amount of sunshine received by any particular point on the planet changes in a periodic manner. The end result is that the Earth goes through a cold snap every 100,000 years or so.[3]

The Earth's orbit has breathed, and wobbled, and tilted, in much the same way for millions upon millions of years, and the general effect has been very small. Or it was – until about two and a half million years ago. Up until then, facts on the ground had been of much greater significance to living things. Matters such as the coalescence and break-up of continents, with the attendant disruption to the chemistry of the oceans and atmosphere. It so happened, though, that two and a half million years ago, the impact of the celestial clockwork above was amplified, rather than dissipated, by the lie of the land below.

With ice already at the poles, the conditions were just right. Cosmic clockwork and continental drift worked together, tipping the entire planet into a series of ice ages. They began gently, but became, in general, more severe, and continue to the present day. Each glacial episode lasts about 100,000 years, with a respite of 10,000–20,000 years or so when the climate can become, briefly, very warm, and even tropical, even at high latitudes.

The coldest part of the most recent cold snap was 26,000 years ago. Much of north-eastern North America was buried under what is known as the Laurentide Ice Sheet, western North America under the Cordilleran Ice Sheet. Most of north-western Europe languished beneath the Scandinavian Ice Sheet. Mountain ranges from the Alps to the Andes groaned under glaciers. Much of the rest of the unglaciated northern hemisphere was a mixture of dry steppe and tundra, treeless and windswept.

All that water locked up in ice had to come from somewhere: the mean sea level was 120 metres lower than it is today. We are currently 10,000 years into a warm spell, and mean sea level is rather higher, on average, than it has been for 2 million years or so.

The changes of climate imposed by ice ages were often very rapid, and have been, to say the least, disruptive. The greatest contrasts can be seen in Britain, which lies at the extreme western edge of the Eurasian landmass, and so is highly sensitive to changes in the ocean, and the prevailing westerly winds. Half a million years ago, Britain was buried under ice a mile thick. In contrast, the climate was so warm 125,000 years ago that lions hunted deer on the banks of the Thames, and hippos wallowed as far north as the River Tees. Forty-five thousand years ago, Britain was a treeless steppe where reindeer roamed in winter and bison in summer.[4] Twenty-six thousand years ago, it was too cold even for reindeer.[5]

These dislocatingly abrupt changes in climate have been further modulated by ocean currents, and even the presence of the ice itself.

The main reason why Britain has a mild climate today, especially given its relatively northerly latitude, is that it is bathed in a current of warm ocean water that makes its way north-eastwards from Bermuda, or thereabouts. When this current reaches the general region of Greenland, it meets polar water from the north, cools, surrenders its warm air to the atmosphere, and – because cold water is denser than warm water – sinks to the bottom and moves southwards again, forming part of a worldwide deep-ocean current system.

The climate of Britain is acutely sensitive to the latitude at which the northbound current cools and sinks. Were this current to run much further south than it does now, the climate of Britain would be a great deal colder. During the chilliest parts of the ice ages, the current didn't get much further north than Spain. As a result, the climate of Britain was more like that of northern Labrador than its present equable state.

The worldwide deep-ocean current is driven not just by heat, but also by salinity. The saltier the water in the warm, north-east-trending current in the North Atlantic, the denser it will be, and the more smartly it will sink to the bottom when it reaches Greenland. A side effect of this is that ice, which floats, tends to be less salty than the sea in general.[6]

A problem arose towards the end of the last glacial episode, when a generally warming trend resulted in the calving of icebergs from the Laurentide Ice Sheet into the North Atlantic. The sudden dumping into the sea of enormous quantities of cold, fresh water made the sea less salty, so the turnover of water into the deep oceans was weakened.[7] The result was a series of short cold snaps, within the general warming trend.

As for the ice itself, it is very bright, and reflects sunlight. The more ice there is, the more sunlight is reflected back into space, the less the ground is warmed, which leaves more ice unmelted, which reflects more sunlight; and so it goes, in a positive feedback loop.

All these factors mean that the effects of the majestic

celestial clockwork are less perfectly predictable than one might imagine, and climate change can be very sudden. At the very end of the last glaciation, about 10,000 years ago, the climate of Europe went from subarctic to equably temperate in the space of a human lifetime.

The dramatic shifts in climate were most severe on continental margins, and towards the poles, but their effects were also felt in the tropics, where the various species of hominins were living, if precariously, on the savannahs and forest margins of Africa. The very idea of ice sheets had yet to trouble their darkest dreams. Their immediate problem was that the climate, already dry, became more arid still.

And it all happened, rather suddenly, around two and a half million years ago.[8]

The woods withered.

Game animals became fewer, more skittish, harder to locate and kill.

It was no longer possible for hominins to live a kind of dilettante existence, digging for roots here, scavenging carcasses there. The various species of *Paranthropus* continued doggedly to dig, crushing nuts to splinters and tubers to mush in their powerful jaws, but life for them only got harder. The time came when the roaming groups of *Paranthropus* became rare, and, sometime around half a million years ago, when northern Europe and North America groaned under the heaviest weight of ice yet, they vanished from the savannah.

But at that time a new hominin appeared, very different

from anything that had come before. It stood taller than any hominin had yet. It was brainier. It took the bipedal stance that hominins had adopted millions of years before, and perfected it. Whereas *Paranthropus* had become a specialist vegetarian, and other hominins opportunist gatherers and scavengers, this new breed had evolved to be a savannah predator.

Our name for this creature is *Homo erectus*.

Compared with the hominins that had gone before, *Homo erectus* was built on an altogether different chassis. As the name suggests, it stood that much taller, more upright. Its hips were narrower, and its legs were proportionately longer, making walking more efficient. Its arms were proportionately shorter: climbing was much less important in its daily round. Although hominins had been bipedal for 6 million years, they had always retained some skill in the trees. *Homo erectus* was the first hominin to commit to the two-legged life entirely.

This commitment brought a host of other changes. *Homo erectus* consumed much more meat in its diet. As we have seen, meat is more digestible than vegetable matter, and contains more available nutrients and calories. *Homo erectus* had a smaller gut, and could afford to have a larger brain. This last is important, for brains are expensive to run. A brain comprises a fiftieth of the body's mass but consumes one-sixth of all the available energy.

Because of the smaller gut, *Homo erectus* had a more definite waistline than its somewhat squat, pot-bellied ancestors. Its hips were higher and narrower, allowing the torso to twist easily, relative to the legs. At the same time, it held its head higher, on a much more defined neck. This meant that *Homo erectus* could do something new: it could *run*, swinging its arms in the opposite sense to the strides of its legs, while keeping its eyes and head directed forward, towards its goal.

Running became very important. Although *Homo erectus* was a poor sprinter compared with, say, a cheetah, or an impala, it excelled at endurance running. By being very patient, *Homo erectus* could pursue large prey animals for kilometre after kilometre, hour after hour, until the quarry literally collapsed from heat exhaustion.[9]

The hunters felt the heat much less than their prey. This was, in part, because *Homo erectus* had become much less furry than most other mammals. That is, it had the same amount of hair, but it was fine, and very short. The spaces between were filled with sweat glands that shed water and cooled the body through evaporation – something that furrier animals could not do.

Despite these impressive feats, it took more than one spindly hairless hunter to subdue an antelope at bay, even one on the point of death. More than at any other point in hominin history, it was important for the hunters to work together, in groups.

But the cohesion that was vital at the kill was created at home.

Homo erectus, like many open-country predators, such as hunting dogs, was a social animal. It was given to activities such as sexual display, extreme violence, and cookery.

At some point in its evolution, various tribes of *Homo erectus* learned how to use fire. They discovered, in cookery, a tasty and a sociable experience. They were less aware, at the time, that cooking food released more nutrients and killed any parasites or diseases that uncooked fare might contain. The tribes[10] that used fire lived longer, healthier lives and produced more offspring than those that did not. Eventually, those tribes that did not use fire died out.

The existence of tribes meant that *Homo erectus* was, to some extent, territorial. Primates, more than any other mammals, are prone to aggressive violence, even murder.[11] Hominins are the most murderous of all. But hominins are lovers as much as they are fighters, part of a syndrome that includes social structure, sexual and social display – and the relative hairlessness of hot-weather hunters.

Hairlessness allows for much more than the shedding of heat. Along with a bipedal stance, it also exposes a human being's more tender parts to general view. Public sexual display may account for the otherwise puzzling fact that human males have much larger penises, in relation to body mass, than other apes.

Sexual display – and the need for group cohesion – may also explain why the breasts of human females are prominent at all times, not just during nursing. In other mammals, the teats wither away to virtually nothing when a female is not lactating.

By the same token, the genitalia of human females look the same no matter whether females are ovulating or not. In other primates, a female's external genitalia are often grossly swollen during oestrus, making her reproductive status absolutely clear to any member of the group. In humans, the reproductive status of a female is hidden to such an extent that it is often a secret to the female herself.

In humans, there is no such thing as a 'mating season', during which time, in other mammals, males and females have sex in full public view. This is, in part, a way to demonstrate and enforce social standing. Humans, in contrast, may be fertile (or not) at any time of year, and prefer to have sex when other members of the group are not watching.

Although humans are highly social and sociable, they tend to form stable pair bonds for the rearing of offspring. Although mating systems vary hugely among humans, the general rule is that one male and one female form a bond that lasts for the many years it takes to raise children.

This is reflected in the relatively limited degree of physical difference between males and females – what is known as sexual dimorphism. In animal species where males tend to monopolize a large group of females, males are very much more massive than females. This is true today in the gorilla, an ape that lives in small groups in which a harem of small females is dominated by a single large male.[12] Human males tend on average to be more massive than human females, but this difference is relatively small. In humans, sexual dimorphism is much less about mass than the distribution of body hair and subcutaneous fat.

If humans form stable pair bonds, then, why do human males have such large penises, and why are the breasts of females always prominent, as if individuals of both sexes are always advertising their availability? Conversely, why are female genitalia always modest, irrespective of reproductive status? Why is oestrus always hidden, while sex goes on in private? If pair bonds were fully stable, none of this ought to matter.

The answer is that although couples are best for the immediate raising of offspring, humans indulge in adultery much more than is generally appreciated. It is said that it takes a village to raise a child, and this is especially true of hominin children, which are born in a relatively helpless, underdeveloped state.

Cooperation between families will be favoured if nobody can be entirely sure of the paternity of any particular child. This cooperation will carry over to the camaraderie of the males in any hunting party. Unsure of which child belongs to which father, males hunt not just for their immediate family unit, but the entire tribe.

In many respects, the social and sexual mores of humans have more in common with those of birds than of other primates. Many birds are social, territorial, indulge in sexual display, and live in family groups in which older offspring assist the parents in the raising of younger siblings before leaving home and seeking territories for themselves. Many bird species form pair bonds that are stable in public, but females are not above mating, in secret, with other males, when their nominal partner is away hunting. This means that a male can never be sure which of the offspring he is helping to raise are his own, and which have been fathered by another.[13]

Faced with such a situation, males tend to hedge their bets. In human societies, the best strategy is to cooperate with other males. Adultery, in the end, contributes to male bonding, and keeps societies bound together, despite the appearance of pair bonding.

Homo erectus was much like us. But similarities can be deceptive. If we looked into the eyes of *Homo erectus*, we would not see the shock of recognition, only the cunning of a predator, like a hyena or a lion.[14] *Homo erectus* was disconcertingly inhuman.

Most mammals are born, grow up rapidly, reproduce as quickly as possible, and, as soon as their capacity for reproduction is spent, they die. The same was true for *Homo erectus*. Their young grew rapidly from infancy to maturity without the lengthy period of childhood that characterizes human beings.[15] When they died, their bodies were ignored, left as so much carrion. *Homo erectus* lacked any concept of the afterlife. They had no visions of heaven. They feared no hell. Most importantly, they had no grandmothers to tell them stories and act as reservoirs of tradition.

And yet, and yet – *Homo erectus* was the author of the most beautiful artefacts: those beautiful, expertly worked, tear-drop-shaped, almost jewel-like stones known popularly as hand axes, the signature artefact in their stone-tool culture, the Acheulean.[16]

The hand axe is so distinctive because it has more or less the same design wherever it is found, irrespective of its age or the material from which it is made. Its association with a particular species – *Homo erectus* – suggests that hand axes, for all their undeniable beauty, were made according to a hard-wired, stereotypical design. They were created as thoughtlessly as birds make their nests. If, when creating a hand axe, the maker made a mistake in the sequence of strokes required to chip it from a blank flint, they would not try to fix it, or perhaps turn it to some other purpose. They would simply discard the mistake and start again from the beginning, with a fresh blank.

This – to us – chilling inhumanity is underlined by the fact that no modern human has entirely worked out what hand axes were *for*. Although many are about the right size to fit comfortably in the hand, where they could be used as choppers, some are far too large for such a use. In any case, why bother? It's very easy to strike an edge from a flint that's sharp enough to, say, skin a carcass, or remove flesh from bones. Why, then, go to the trouble to make something as complex and beautiful as a hand axe for the purpose? If one is going to throw stones – or even use a slingshot – to bring down prey or an enemy, why go to the trouble of creating a hand axe, if it is simply going to be thrown away?

We tend to think that items of technology have a purpose, and that this should be evident from their design. 'To see a thing one has to comprehend it,' wrote Jorge Luis Borges in his short horror story *There Are More Things*:

> An armchair presupposes the human body, its joints and limbs; a pair of scissors, the act of cutting. What can be said of a lamp or a car? The savage cannot comprehend the missionary's Bible; the passenger does not see the same rigging as the sailors. If we really saw the world, maybe we would understand it.[17]

Our conceit comes from our tendency to attach to the elaborate construction of objects outside the body a kind of conscious direction or purpose that is distinctly and uniquely human. A glance at a beehive, or termite mound, or a bird's nest, will show instantly that this equation is false.

On the other hand, *Homo erectus* did, on occasion, do what appear to us some very human-like things – such as scratch out hash-marks on seashells.[18] For what purpose, nobody knows. It's also possible that *Homo erectus* mastered the art of sailing, or canoeing, into the open sea – as human an urge as can be imagined. As we have seen, *Homo erectus* was the first hominin that learned to tame and use fire.

Whatever else it was, or did, or thought, *Homo erectus* was one of evolution's responses to the sudden shift in climate about two and a half million years ago. Rather than retreating, as the remaining apes did, to the dwindling forests, there to live a kind of theme-park existence as a memorial to a vanished past;[19] or trying to prise an ever-more-meagre existence from the hard-scrabble savannah, as *Paranthropus* tried to do, and eventually failed; *Homo erectus* began to range more widely than other hominins had done, just to eke out a living from the unforgiving Earth.

Eventually, *Homo erectus* was the first hominin to leave Africa.

By 2 million years ago, *Homo erectus* had spread across the continent.[20] But it didn't let the savannah grass grow under its feet. As a result of the change in climate, the forests had shrunk to such an extent that savannah rolled unbroken across Africa, the Middle East, and Central and Eastern

Asia. The endless grassland heaved with game, and *Homo erectus* followed it, wherever it led.

It was already pursuing the herds as far as China as long as 1.7 million years ago, and perhaps even earlier.[21] Three-quarter of a million years ago, *Homo erectus* made regular use of caves at Zhoukoudian, now in the suburbs of Beijing.[22]

And as *Homo erectus* spread, it evolved.

Homo erectus was the protean progenitor[23] of a huge variety of daughter species: comparable to giants, and hobbits, and troglodytes, and yetis, and – ultimately – to us. The tendency towards variegation started early. One tribe of *Homo erectus* that lived in Georgia, in the Caucasus Mountains, some 1.7 million years ago, was of such motley cast that it is hard, from our modern perspective, to imagine that they all belonged to the same species.[24]

By a million and a half years ago, tribes of *Homo erectus* had penetrated island South East Asia. To do this they need only have walked. The sea level stood so low that most of the region was dry land. The many islands we see today are the half-drowned fragments of a region once much vaster in extent. *Homo erectus* lived in Java until at least 100,000 years ago[25] – the last holdouts clinging on, as the sea level rose and jungle once again pressed in all around them.

They may even have survived long enough to witness the arrival in the region of their descendants – modern humans.[26] If they met, the encounter would have gone badly for what, to modern humans, would have seemed like a large but secretive woodland ape, just one of several native to the region, such as the orangutan and its enormous cousin *Gigantopithecus*.

Once within island South East Asia, the evolution of *Homo erectus* took some surprising turns. Confined to islands as the sea level rose, various tribes, sundered from the mainland, evolved each in its own peculiar way.

One such reached Luzon, in the Philippines, where it hunted the native rhinoceros[27] at about the same time as its mainland cousins were kindling sparks in eastern China. Once marooned, these people evolved into *Homo luzonensis*, a species of tiny size.[28] As well as being small, they were, in many respects, primitive. With the return of jungle, these hominins once again took to life in the trees. They survived until at least 50,000 years ago. When the first modern humans arrived, these atypical descendants of an African savannah hunter must have stared down from the branches at the new invaders with incomprehension and horror.

As strange a fate awaited another group of *Homo erectus* that reached Flores, an island well to the east of Java.

They made landfall more than a million years ago. This is in itself surprising, as they could not simply have walked there, as their forebears had to other islands closer to the continent. Even when the sea level was at its lowest, Flores was separated from the rest of the world by deep channels.

It is possible that they got there by accident, perhaps having been blown on wings of storm, cast ashore by a tsunami caused by an earthquake or volcanic eruption, and rafting on vegetation or other debris. This part of the world is, after all, no stranger to extreme events, and such accidents explain the existence on even the most remote islands of plants and animals.

Either that, or they reached Flores by some form of watercraft, even if that watercraft was intended for fishing close inshore to another island and got blown off course.

However they got there, when they reached Flores, they too shrank in size over time,[29] and became what we know as *Homo floresiensis*. By the time they became extinct, around 50,000 years ago, about the same time as their distant cousins in the Philippines,[30] they stood no more than a metre in height, but made tools as well as their ancestors had done, if on a smaller scale, to fit smaller hands.

This miniaturization is not unusual; strange things happen to species that become marooned on islands. Smaller animals evolve to be larger, and large animals evolve to be smaller.

The monitor lizards of Flores, cousins of the Komodo dragon, evolved to a size that would have been truly frightening to a modern human, let alone a person of a metre in height, no matter how dauntless. Some of the rats evolved to the size of terriers.[31]

As a consequence of the frequent rise and fall of the ice-age oceans, many islands could boast their own unique species of tiny elephant, and Flores was no exception. Perhaps *Homo erectus* came to Flores in search of large elephants, and, over the millennia, both hunter and hunted became smaller as each adapted to island life.[32]

Even accounting for its small size, *Homo floresiensis* had a very tiny brain. But as the savannah hominins had discovered when they had become carnivores in Africa long before, brain tissue is notoriously expensive to maintain. In a species challenged by scarcity to the extent that dwarfism might be favoured by natural selection, there is even more pressure on brains to do more with less. Smaller brain volume alone need not compromise intelligence: among the birds, crows and parrots are notoriously clever despite having brains no larger than nuts. *Homo floresiensis* made tools neither more nor less sophisticated than those of *Homo erectus*.

On Flores, Luzon, and almost certainly elsewhere, *Homo erectus*, once marooned, became smaller, turning into what we might see as dwarves or hobbits.

Elsewhere, they became giants.

In Western Europe, the species transmuted into *Homo antecessor*, a rugged creature that ranged well outside the warm savannah of its ancestor. Some 800,000 years ago, it left hand axes, and even footprints, in eastern England – much further north than any hominin had yet ventured.[33] Hardy, but strangely familiar, *Homo antecessor* looked much more like a modern human than *Homo erectus*, or even that acme of ice-age cave life, the Neanderthals. Our human physiognomy has deep roots, as do our genes: it is in *Homo antecessor* that the first signs of genetic kinship with modern humans can be found.[34]

Somewhat later, and elsewhere in Europe, appeared *Homo heidelbergensis*. The bones and tools that have come down to us from their European heartland show that they were formidable indeed. Their hunting javelins, preserved in Germany along with stone tools and the butchered remains of horses, and dating to about 400,000 years ago, appear to us more like fence posts.[35] These spears – one of them is 2.3 metres long and almost 5 centimetres in diameter at the widest point – were designed not to be thrust, but thrown. To have lifted and used these weapons in battle must have required great strength. A shin bone from southern England[36] is similar in size to that of a modern adult human male, but much denser and thicker, indicating an exceptionally robust individual that weighed more than 80 kilograms. At the other end of Eurasia, humans comparable in size with only the tallest of modern people strode out of the snows of Manchuria. There were giants on the Earth in those days.

The descendants of *Homo erectus* in Europe and Asia were, clearly, evolving in response to the ever-harsher conditions of the ice age. The slender long-distance runner of the African savannah was turning into something new, different – a creature tough enough for the rigours of the north.

About 430,000 years ago, a tribe settled in caves in the Sierra de Atapuerca[37] in northern Spain. In many ways they looked human. Their brains were around the same size as those of modern humans. But their faces were heavily set, tough. Their outlook on a bleak world was offset by a deepening inner life. For they buried their dead. At least, they did not let them lie unmarked, as if dead bodies were any other object: the bodies were taken to the rear of the cave, and thrown into a deep pit. In these people were the beginnings of the Neanderthals.[38]

The Neanderthals, perhaps even more than *Homo erectus*,

exemplify how life evolves in response to environmental challenges. Thrawn, supremely adapted to life in the cold, windswept wastes of northern Europe, they lived there unchallenged for 300,000 years. They rode light on the landscape, their culture changing little. But with brains on average larger than that of a modern human, they were thoughtful and deep. And they buried their dead.

Deep in caves, far from the cold, the wind, and the weak sunlight of the ice age, they strove for the numinous. In a cave in France, buried so far underground that sunlight could never have penetrated, they built circular structures of smashed stalactites and the bones of bears.[39] For what reason, nobody knows. These mystifying structures are 176,000 years old. They are the oldest well-dated constructions made by any hominin.

Neanderthals made a stark contrast to their lithe, free-ranging *Homo erectus* ancestors. Although they and their works have been found from the western extremes of Europe through the Middle East and into southern Siberia, individual Neanderthal groups did not range far across the landscape. Faced, externally, with extremes of climate no hominin had ever experienced, they made brief sorties outdoors for food, but cultivated a brighter life of the mind – like H. G. Wells's Morlocks – under the Earth.

Some of their relatives, however, set their sights even higher.

Some time before 300,000 years ago, an offshoot of the Neanderthals in central Asia looked up and saw the Tibetan Plateau. Outside the polar regions, this is perhaps the least hospitable part of the world for humans. The air is cold, harsh and thin. The snows never melt. When the sun shines, it is a searing eye in the ice-blue vault. Yet a group of hominins felt that, high on the Roof of the World, they could scratch a living. And so they did. They climbed. And, as they climbed, they evolved. They turned into the Denisovans,[40] reminiscent of the yetis who according to legend inhabited the plateau thousands of years later.[41]

Homo erectus and its descendants conquered the Old World. They might even have ventured into the New.[42] Around 50,000 years ago, many species of humans walked the Earth. There were Neanderthals in Europe and Asia. By that time, some of the descendants of the Denisovans had left their mountain fastnesses and descended as far as the highlands of eastern Asia.[43] Everywhere they went, they changed to meet the challenges of new environments, from deep caves to tree-tangled jungles to isolated islands to open plains to the highest mountains. *Homo erectus* itself was still peacefully living in Java.

And yet, all these experiments in human life were to be swept away. By the end of the ice age, only one species of hominin was left. Like *Homo erectus*, it came out of Africa.

Timeline 6. *Homo sapiens*

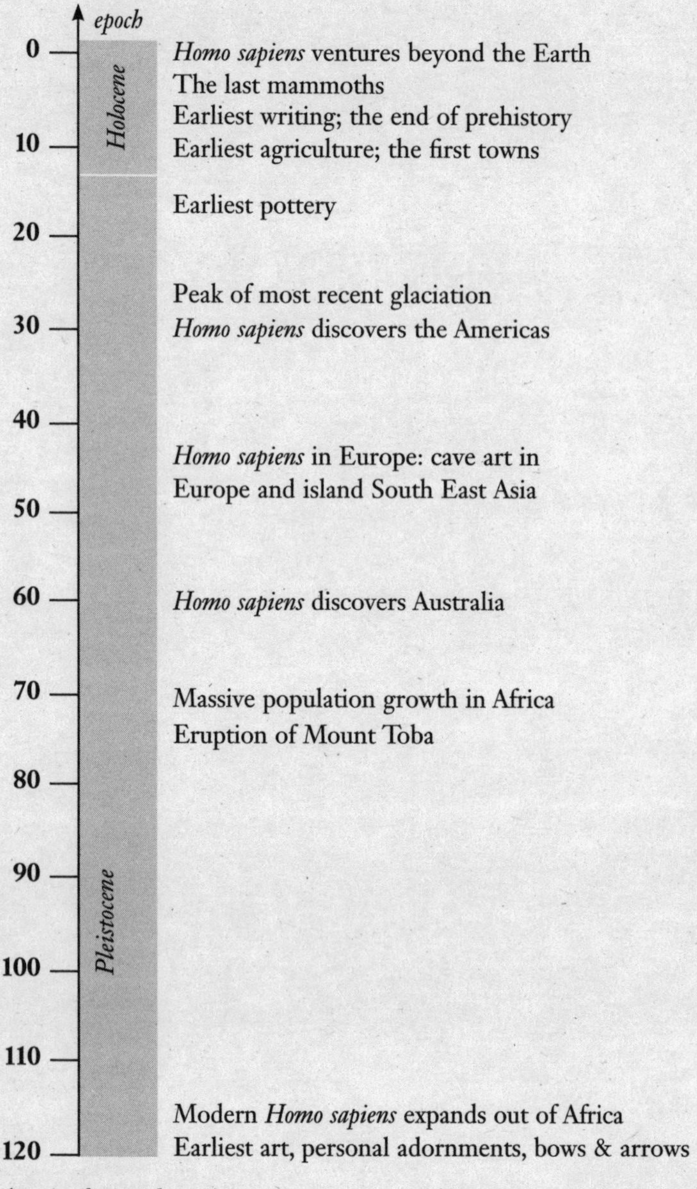

epoch
0
10
20
30
40
50
60
70
80
90
100
110
120
Holocene
Pleistocene
Homo sapiens ventures beyond the Earth
The last mammoths
Earliest writing; the end of prehistory
Earliest agriculture; the first towns
Earliest pottery
Peak of most recent glaciation
Homo sapiens discovers the Americas
Homo sapiens in Europe: cave art in Europe and island South East Asia
Homo sapiens discovers Australia
Massive population growth in Africa
Eruption of Mount Toba
Modern *Homo sapiens* expands out of Africa
Earliest art, personal adornments, bows & arrows
Ages in thousands of years before the present

11

THE END *of* PREHISTORY

By around 700,000 years ago, glacial episodes were very much longer than the warm intervals that separated them. The Earth was now in a more or less permanent state of glaciation. Respites were hot, heady and brief.

Life not only survived – it thrived. Those parts of Eurasia that did not labour under the ice were clothed in green steppe that supported an almost incalculable tonnage of game. Spring and summer saw bison migrating across the land in herds so huge that it would have taken days to watch the millions of animals pass by. These herds were joined by horses, and giant deer with incredible spreads of antlers; punctuated by elephant species such as mammoths and mastodons; and accompanied by the snort and stamp of woolly rhinoceroses. Winters were only slightly less abundant. Although many of the animals migrated southward, the snows were accompanied by reindeer. All this meat on the move was a magnet for carnivores such as lions, bears, sabre-toothed cats, hyenas, wolves – and the tough, hardy heirs of *Homo erectus*.

Hominins responded to the deepening ice age with larger brains, and more abundant stores of fat.

This was, in itself, remarkable. As we have observed, brains are very costly organs to run. The economics of nature usually demand that a brainy animal lay down only minimal fat stores, because if food runs low, it will have the cunning to find more before it starves to death. It's only the dimmer flames among mammals that need to lay down fat. Humans, though, are the exception.[1] Even the skinniest humans store much more fat than the plumpest apes. Brainy animals that have a good layer of insulation have all they need to cope with the endless cold of the ice age.

Fat had another purpose, too. The difference between the sexes is largely a matter of fat deposition. The body of an adult human male contains, on average, about 16 per cent fat by weight; of a female, 23 per cent. This difference is significant. Onboard energy is an essential prerequisite for fertility and pregnancy, particularly in times of dearth. As such, the mechanics of selection favoured those females with plump, rounded curves as having the best prospects for reproduction.[2]

Big brains, though, can also pose problems. Big brains mean big heads. Human babies, with their big heads, have a hard time being born. Babies are only born thanks to a ninety-degree twist of the head as they negotiate the mother's pelvis and emerge from the vagina. Until very recently, the cost was borne by the mother, who stood a high risk of dying in the process. Human babies appear in a relatively helpless state. If they waited until they were more developed, and perhaps better able to cope with the world, they might be too big to negotiate the birth canal, and not get born at all. The nine months of pregnancy represent an uneasy truce between the baby, which needs to be as able to cope on its own in the outside world as soon as possible, and the mother, who, if she waited any longer, would have to play increasingly loaded dice with death.

It is a compromise that suits nobody. A species in which babies are born utterly helpless, and, even if born successfully – to mothers that stand a high risk of death – take many years to raise to maturity, is one that is likely to become extinct very quickly. The solution was a dramatic change, but at the other end of life. That change was the menopause.

The menopause is another evolutionary innovation unique to humans. In general, any creature, mammalian or otherwise, that is too old to reproduce will age and die very quickly. In humans, though, females that have ceased the capacity to reproduce in their middle years can expect to enjoy many decades of useful life – and, in the end, raise more children.

The enlargement of the brain and the consequent helplessness of babies was accompanied by the emergence of grandmothers:[3] post-menopausal women who would be there to help their daughters raise their grandchildren. The logic of natural selection says nothing about who actually raises children to maturity – just as long as they are raised by somebody. It so happens that a woman who ceases reproduction in order to help her daughters raise grandchildren will raise a greater number of descendants, on average, than had she remained reproductive herself, competing for resources with her daughters. Over time, those groups of humans that could rely on post-menopausal women to help raise children would raise more of those children to reproductive age themselves. Those unable to tap such a valuable resource died out. Uneasy compromise was trumped by cooperation.

Reproduction takes energy away from everything else. There is, in general, a trade-off between reproduction and longevity. So, by ceasing reproduction in middle age, human females actually *increased* their reproductive output – *and* lived longer. The enlargement of the brain led to increased life expectancy at birth, from perhaps the mid-twenties in *Homo erectus* to the forties in Neanderthals and modern humans.

Even though the pressures of evolution acted differently on males and females, they shared the same genes, and this led in effect to a war between the sexes as pressure was brought to bear on genes by selective forces acting in opposition – one gene, two masters. The result was another compromise. Because females had to be fatter to bring babies into the world, males became fatter too, but not as much. Because females evolved the menopause and lived longer, males came to live longer too, but not as much.[4] The result was the introduction of a new stratum in hominin society – the Elders, of both sexes. Before the invention of writing, the Elders came to be valued as the repositories of knowledge, wisdom, history, and story.

For the first time in evolution, there existed species in which knowledge could be transmitted across more than one generation at a time. Many animals are capable of learning. Whales learn their songs from other whales; birds their songs from other birds; puppies the rules of play from other puppies; human babies, language by the unconscious imitation of the humans around them. Humans are unique, as far as is known, in that they are the only animals that not only learn, but teach.[5] The Elders made this possible. While younger members of the tribe were nursing babies or out hunting, the less immediately productive Elders were passing on their stores of knowledge to new generations – of children who, with their long childhoods (a function of their relative immaturity at birth), had plenty of time to acquire this knowledge. Abstract information became a currency for survival as important as calories. The consequences were to be explosive. And it all began during the ice age, when, for the first time, it was an advantage for a primate both to store fat, and have a larger brain.

The deepening cold of Eurasia was matched by the aridity of Africa. The sere savannah graded into dry desert, punctuated by waterholes as evanescent as mirages. Survival was a constant struggle. Extra onboard storage of fat was an advantage here too, just as it was closer to the ice sheets. Humans adapted by evolving a metabolism based on boom and bust – they were able to go for many days without food, but, when they made a kill, could gorge themselves to the threshold of pain, until they could literally not eat another morsel, or even move – all the better to absorb as many nutrients as they might need to survive until the next meal, whenever that might be. Humans ate with gusto, as if any meal might be their last.[6]

Despite the constant threat of extinction – and perhaps even because of it – the heirs of *Homo erectus* diversified in

Africa, as they had everywhere else.[7] Then, a little over 300,000 years ago – just as the first Neanderthals were adapting to the icy chill of Europe – a new hominin appeared in Africa. It was rare, variegated and scattered, but ranged all across the continent.[8] To have met these individuals would have been to look ourselves in the face. These were the very first of our own species, *Homo sapiens*.

Behind the face, though, these new creatures were not quite as human as they first appeared. At first, *Homo sapiens* was something of a raw ingredient. Modern humans were to be hardened by more than a quarter-million years of failure. For the first 98 per cent of its existence, the tale of *Homo sapiens* was one of heartbreaking tragedy, had the participants survived to tell the tale. Almost all, however, perished, and the species almost died out completely.

Along its journey, however, its gene pool acquired a seasoning of DNA from other hominins, both within Africa and beyond it. *Homo sapiens* is a species with many parents, each of which added its own special flavour to a mix that eventually succeeded – against all odds.

Even at the beginning, *Homo sapiens* ranged outside its African heartland, making forays into southern Europe around 200,000 years ago, and the Levant between 180,000 and 100,000 years ago.[9] But these excursions left little trace, like the stain of water on desert sand. *Homo sapiens* was still a tropical species, a fair-weather visitor. If the conditions in Africa were harsh, those in Eurasia were harsher. And, were *Homo sapiens* to have persisted, it would have found that the way was shut: barred by the Neanderthals, which, in their heyday, were vastly more cultured, and, used to the long cold of Europe, could afford to play a long game. They would have noticed humans, if at all, as no more than occasional visitors, like a light frost before dawn on a summer's day.

Matters were hardly better for the new species in its African heartland. Indeed, as the ice ages wore on, conditions became steadily worse. The bands of *Homo sapiens*, never very common to begin with, faded out, first in one place and then in another, either dying out, or interbreeding with other varieties of hominin before these hybrids, too, vanished. The time came when *Homo sapiens* had all but disappeared north of the Zambezi. At the last, *Homo sapiens* became confined to an oasis in the north-western edge of what is now the Kalahari Desert, just to the east of the Okavango Delta.

Early in the ice age, the region had been lush. The area was watered by Lake Makgadikgadi, which, at its greatest extent, was the size of Switzerland. As Africa continued to dry out, the lake fragmented into a landscape of smaller lakes, watercourses, wetlands and woods, where giraffe and zebra roamed.

The last ragged remnants of *Homo sapiens* found refuge among the ponds and reed beds of the Makgadikgadi wetland around 200,000 years ago, rather in the same way that, millennia later, King Alfred's last redoubt was in the Athelney marshes – where he regrouped, sought solace, burned a few cakes, and emerged to defeat the Danes and take back the Kingdom of Wessex. If England began at Athelney, the roots of humanity itself may well lie in the

Makgadikgadi wetland. If there was ever a Garden of Eden anywhere, it was there.[10]

Rather in the manner of the ugly duckling, *Homo sapiens* hid in the Makgadikgadi wetland for 70,000 years. But when it finally emerged, it had become a swan.

For tens of thousands of years, the Makgadikgadi wetland was an oasis surrounded by increasingly inhospitable terrain – dry desert and salt pans. Once *Homo sapiens* had settled there, it was far from straightforward to leave. Then, around 130,000 years ago, the Sun began to shine somewhat more brightly on the Earth than it had for some time. The celestial clockwork of eccentricity, axial tilt and precession had contrived to produce an interval of rather warmer climate than the planet had seen for many millennia.

In Europe, the great glaciers were replaced – albeit briefly – by near-tropical conditions. This was the epoch when, in Britain, lions gambolled in Trafalgar Square, elephants grazed in Cambridge, and hippos wallowed where the city of Sunderland now stands. As in Britain, so in Africa – the climate mellowed. The latest generations of *Homo sapiens* found that the desert beyond the Makgadikgadi had become a sea of grass.

They moved out, following the game – and in time entirely so, for, before long, the Makgadikgadi dried out completely. Today it is a salt desert that supports no living thing more complex than crusts of cyanobacteria, a throwback to the earliest days of life on Earth.

The bands of *Homo sapiens* tracked the game south, until they reached the coast at the southernmost extent of Africa. When they did, they developed a whole new mode of life, based on the protein-rich abundance of the sea. For people used to scratching an existence on tough roots, unpredictable fruits and the skittish habits of wary game, the ocean was a feast beyond imagining. Shellfish, bursting with protein and essential nutrients, and totally unable to run away. Tasty and salty seaweed, and fish, much easier to trap than impala or gazelle.

As if they were breathing a collective sigh of relief after their long history of adversity, these early beachcombers became much more settled and started to do things that humans had never done before. Feasting, they garlanded one another with necklaces of shell beads. They painted themselves with charcoal and red ochre.[11] They engraved their signs as cross-hatched patterns on ostrich egg shells, and painted them in ochre on rocks.[12] To be sure, Neanderthals and even *Homo erectus* occasionally used shells and made engravings, but these people were engaging in such activity with a new intensity and commitment.

At first, such technologies seem to have appeared and disappeared like will-o'-the-wisps, as if humans occasionally lost the knack or the inclination. But the use of technology deepened and became more habitual as the population slowly rose and its traditions cemented. These beach-dwellers also started using stone in a new way. Rather than chipping away at rocks to make artefacts that could fit in a fist, they created much smaller, carefully crafted, heat-hardened pieces that could be, for example, hafted onto arrows. They invented projectile weapons. Weapons that could kill game at a distance, with relatively little risk to the assailant.[13]

Other exiles from Eden made their way in the opposite direction, to the north. The Zambezi was their Rubicon. Once they reached East Africa, they were joined by emigrants from southernmost Africa, who introduced their advanced technologies – their cosmetics, their shell necklaces, and, most of all, their bows and arrows. The result was explosive. The population of *Homo sapiens* in East Africa expanded from a few small bands into something resembling a population that stood a chance of more than a mayfly existence.[14] By about 110,000 years ago, they had spread throughout Africa once more and were making renewed steps outside their homeland.

And this time, they would come to stay.

It came like a firebell in the night. Around 74,000 years ago, a volcano called Mount Toba on the island of Sumatra erupted explosively, in an event as catastrophic as anything that had happened on Earth for millions of years.[15] It brought the period of relative warmth, already decaying, to an abrupt close. Debris rained down over the entire Indian Ocean region, even as far away as the coast of South Africa.[16] Hundreds of cubic kilometres of ash were lofted into the atmosphere, plunging the world into sudden glacial chill.

In an earlier age the disaster might have scrubbed nascent humanity completely from the face of the Earth. This time, *Homo sapiens* seems barely to have paused. By then, our species had spread from Africa around the Indian

Ocean basin. Flint-knapping humans were in India,[17] had roamed as far as Sumatra itself[18] – the epicentre of the blast – and had reached southern China.

When the exiles from Makgadikgadi left their oasis, the first thing they did was head for the beach. When later humans left Africa, they did so, at first, by following the shoreline, across southern Arabia and India and into South East Asia. They also travelled inland, along the courses of rivers, and into the savannah, when the climate permitted.

We should not think of the event as a Mosaic exodus: more a series of events which, tiny in themselves, combined to create only what looks like a predetermined pattern. What people did not do was look up at the horizon and, in a fit of heroic prolepsis, approach some manifest destiny. Individual humans on the ground lived their entire lives in more or less the same place. Pressure of population suggested that some people might move, say, beyond the next headland. Inclement weather would have forced many reverses. Humans in different but adjacent tribes, joined by interwoven threads of relationship, would meet at times of festival to sing, dance, swap tall tales and choose mates. As in all primates, a female, once matched, would move from the country of her ancestors and set up home with the family of her mate in some far place – across the river, maybe, or over the next hill.[19]

The migration was, therefore, not a single event, but a series of small ones. It did, however, prove to have an overall shape. It pulsed, along with regular climate changes forced by the Earth's orbital cycles; in particular, the 21,000-year cycle of precession.[20] The migrating humans followed their stars – but they would have been different stars, at different times.

As a species, humans seem to have had especially itchy feet between 106,000 and 94,000 years ago, when they spread across once-hospitable southern Arabia and into India; between 89,000 and 73,000 years ago, when they reached island South East Asia; between 59,000 and 47,000 years ago – a particularly busy period of migration across Arabia and into Asia, which also saw landfall in Australia;[21] and, finally, between 45,000 and 29,000 years ago, which saw the thoroughgoing occupation of the whole of Eurasia, including at high latitudes, as well as tentative steps into the Americas – and also some migration back into Africa.

This does not mean that humans stayed still outside these times – these were the intervals when climate was clement enough for migration to be most favourable. There were times when the spreading human population was divided. For example, the cold, dry spell just after the Toba eruption saw African humanity cut off from the population in southern Asia. They were not to meet again for another 10,000 years.

On their way, migrating humans met other hominins. The encounters were rare: their outcome, variable. At times, the tribes, sensing difference, would fight. At others, they would greet one another as cousins from afar, realizing that, after all, they were not as different as they first appeared. They bonded by telling tales and swapping mates. Modern humans met Neanderthals in the Levant and interbred with them. As a result, all modern humans with ancestries that are not exclusively African contain some Neanderthal

DNA.[22] In South East Asia, migrating humans added genes from the Denisovans to the human gene pool – the descendants of the mountain-dwellers, now long since acclimatized to the lowlands. Denisovan genes are now found very far from the mountain fastnesses where they originated, in people in island South East Asia and across the Pacific. But in a curious twist of fate, the gene that allows modern Tibetans to live untroubled in the thin air at the Roof of the World was a parting gift from those people of the Eternal Snows,[23] who themselves disappeared as a separate species after 30,000 years ago – absorbed completely by the greater tide of *Homo sapiens*.

Around 45,000 years ago, modern humans finally broke into Europe, on several fronts, from Bulgaria in the east to Spain and Italy in the west.[24] The Neanderthals, dominant in Europe for a quarter of a million years, had repulsed all earlier incursions from *Homo sapiens*. This time, however, they went into steep decline, and by 40,000 years ago, this acme of the ice age was all but extinct.[25]

The reasons have been much debated. They might have fought with modern humans. They certainly interbred with modern humans.[26] It is possible that they faded out without much of a struggle, in the face of a species that bred ever so slightly faster, and perhaps ranged further from their home territory.[27] In the end, there were so many modern humans in Europe that the remaining Neanderthals, holed up in their last, far-flung redoubts – from southern Spain[28] to Arctic Russia[29] – were too few and widely scattered to be able to find mates of their own kind.[30]

Neanderthal populations had always been small. As they got smaller, the effects of inbreeding and accidents took their toll. There comes a point in any human society when it is too small to be viable. There is nothing that leads a population to extinction as surely as a lack of people.[31] In the end, it was simpler to interbreed with the invaders. DNA from a 40,000-year-old human jawbone from a cave in Romania shows that its owner had had a Neanderthal great-grandparent.[32]

From eastern Europe, modern humans followed the course of the Danube where, at its headwaters, there is evidence of a flowering of cultural exuberance.[33] They made sculptures of animals, humans, humans with animal heads, and even bas-relief ducks that they could hang on suburban cave walls.[34] They made, again and again, sculptures of obese, pregnant and huge-breasted women – poignant invocations of the importance of abundance and fertility in a society that was never far from starvation. They were appeals to a higher power.

Images of animals appeared on cave walls at opposite ends of Eurasia, more or less simultaneously. The justly famous cave paintings of France and Spain have been joined by similar examples in Sulawesi and Borneo, in Indonesia.[35] These, too, had ritual content. Cave art tends to appear in spaces that are acoustically resonant. The pictures seem likely to have been just one component of rituals that also included music and dance.[36]

When human beings came of age, they were invited by a shaman into these ritual spaces for initiation into the tribe. As part of the ceremony, the inductee was painted with ochre, or soot, and told to make an impression of their hand on the cave wall: as if to make their mark in the book of life. To say, 'I am Here.'

After four and a half billion years of mindless tumult, the Earth had birthed a species that had become aware of itself. And what, it wondered, would it do next?

12

THE PAST *of* *the* FUTURE

All happy, thriving species are the same. Each species facing extinction does so in its own way.[1]

As a consequence of climate change, forests are broken down into small spinneys, each stand isolated from the others amid an ocean of grass where once there were trees.

As ice caps melt, the drowning of the land leaves isolated islands where once were mountaintops.

What happens to the life forms clinging to the remnants of what, for them, had been much larger worlds?

Some groups take advantage of this isolation to evolve into strange new forms. One thinks, for example, of *Homo floresiensis* and the dwarf elephants it hunted. Many other isolated populations, however, will find themselves too small to be viable. There might be too little food or water to survive. Individuals will fail to find mates, or, if they do, they might perforce be close relatives, and the population will succumb to inbreeding.[2] Others will simply fail to adapt, trying to live according to old habits in circumstances that have greatly changed.[3] One by one, the individuals die, by genetic indisposition, or by age, or by accident, leaving fewer and fewer offspring, until there are none left. The population has become extinct.

Eventually, when all other populations of the species have failed, each facing its own travails in the fragments of the once-extensive habitat in which it finds itself so marooned, the very last surviving population is at increased risk of succumbing to some very specific, very local disaster. This could be almost anything, and very far from the grand apocalypse of asteroid impacts or the eruption of magma fields. It could be a landslide that extinguishes its only source of food, or something as seemingly prosaic as the bulldozing of its last refuge to make way for a construction project.

Other species may seem abundant, having no reason to fear that their disappearance may be imminent. Closer examination may reveal that they are long overdrawn in the book of life, marked for extinction as surely as if they had been scythed down in their prime. While they may be abundant in the habitat to which they have become accustomed, further removal of habitat – even if moderate – may ensure their extirpation. They are, quite literally, living on borrowed time. For example, the disappearance of butterflies and moths from chalk grassland is better explained by removal of their habitat many decades in the past than current habitat loss.[4] These species have incurred what is called an 'extinction debt'.[5]

Yet other species will, for some reason or other, reduce their rate of reproduction, and the death rate will outpace the rate of replacement.

Homo sapiens has been instrumental in creating

conditions in which many different species have become extinct. By the same token, *Homo sapiens* itself might be vulnerable to one or more of these different means of departure.

Large-scale extinction events in the distant past are so remote that it is difficult to tease out individual stories from the general noise and mêlée of disaster.

The ultimate cause of the mass extinction at the end of the Permian, for example, was the upwelling of lava in Siberia that released gases that raised the temperature of the atmosphere sharply through the greenhouse effect, and poisoned the air and the oceans. But however cataclysmic the event, and however much living things suffered in common, each individual animal or plant, each coral polyp and each pelycosaur, met its death in its own way. Such mass extinctions, therefore, represent the summation of many individual premature deaths, each one its own distinct tragedy.

The end of the Pleistocene, around 10,000 years ago, was marked by the disappearance of virtually all animals more massive than a large dog throughout Eurasia, the Americas and Australia. The ultimate cause of the extinction might have been the spread of rapacious humanity. Alternatively, it might have been dramatic climate change of the kind so often seen during the Pleistocene. Most likely, it was a mixture of both.

However, the end-Pleistocene extinctions are much closer to us in time than the end-Permian catastrophe. The vestiges of the event are fresher, and can be picked over more finely. The fates of individual species can be traced.[6]

For example, the home ranges of two iconic ice-age species – giant deer (popularly known as the 'Irish Elk') and the woolly mammoth – shrank dramatically in just a few thousand years. The precipitous decline was coincident with sudden changes in climate, and the vegetation on which they depended.[7] Hunting, too, will have only hastened a demise that would have happened sooner or later. Giant deer and mammoths might have gone, but their fossils are abundant, and can be reliably dated, so their decline and fall can be mapped in intense detail. Had they departed at the end of the Permian, we could probably say very little more than that they had disappeared, and that would have been that.

More recent extinctions can be dated very precisely. The very last wild ox, or aurochs (*Bos primigenius*) was shot in Poland in 1627. Given the spread of people with guns, this was an extinction that was bound to happen. That said, it represented extinction at its sharpest, most particular and poignant: the single bullet that felled that single ox put an end to the last remaining individual of a species that had once been abundant throughout Europe. In contrast, the northern white rhinoceros (*Ceratotherium simum cottoni*) is, at the time of writing, still with us. Immense efforts are made to ensure that the remaining individuals do not meet oblivion at the point of a marksman's bullet. However, as the population consists of only two individuals, both of them female, it is only a matter of time – and not very much time at that.

The case of the aurochs and the rhino, however, are different. The aurochs belonged to one of the few branches of the mammalian family tree – the bovid family, which includes goats, and sheep, and a legion of antelope species – that is still thriving. Were it not for humanity, the aurochs might still be with us. The rhinoceros, in contrast, had its day back in the Oligocene, when rhinos and other odd-toed ungulates were very diverse, but since when they have been in long-term decline: largely outcompeted by even-toed ungulates such as bovids, of which the aurochs was one. Humanity has only hastened an end that had been all but written long before humans evolved.

The world is currently just 2.5 million years into a series of ice ages that will last for tens of millions of years more. Already, ice has waxed and waned more than twenty times, leading to climatic disruption of a scale not seen since the Eocene. And it's just getting started. With each advance of ice, with each retreat, the game is changed. Some species will die out. Others will flourish. Those that flourish in one cycle might perish in the next.[8] And there will be almost a hundred more glacial-interglacial cycles before the ongoing series of ice ages comes to a close.

Homo sapiens has reaped the benefits of the current cycle. The species jolted into self-awareness as the previous interval of warmth, some 125,000 years ago, decayed into a prolonged cold stage. It took advantage of low sea levels to migrate, hopping between one otherwise isolated island and another.

By the time the ice had advanced to its greatest extent, some 26,000 years ago, humanity had set up camp throughout the Old World, and had even crossed into the New.[9] Only Madagascar, New Zealand, the further Oceanic islands and Antarctica were yet to feel the press of a human foot on their shores – and even they would do so in time.[10] During this advance, all the other species of hominin disappeared. *Homo sapiens* is the very last. The only one left.

For almost all their history, humans were hunters and gatherers, and, like all wise foragers, knew the best places to hunt and gather. Not long after the maximum ice advance, repeated visits to the same spots to harvest useful plants exerted natural selection on these plants to produce fruits and seeds that were most attractive to the visitors. Bakers started grinding the seeds of wild wheat and barley into flour and baking bread at least 23,000 years ago.[11] Agriculture started in several different parts of the world more or less simultaneously at the very end of the Pleistocene, 10,000 years ago.[12]

Since then the rise of the human population has been dramatic. At the moment, this single species consumes a quarter of all the products of plant photosynthesis on Earth.[13] Inevitably, such sequestering means fewer resources for all the millions of other species, some of which are vanishing as a result.

However, most of the rise has been very recent indeed. The exponential growth in human population is a matter of living memory. The population has more than doubled in my lifetime,[14] quadrupled since my grandparents were born. Against the backdrop of geological time, the sudden rise of humanity is of negligible significance.

Most of the impact of humanity on the planet has been felt since the Industrial Revolution, which began some 300 years ago, when *Homo sapiens* harnessed the power of coal on an industrial scale.

Coal is formed from the energy-rich remains of Carboniferous forests. Slightly later, humanity learned how to locate and extract petroleum, an energy-dense mixture of liquid hydrocarbons, created by the slow transformation of the fossils of plankton as they are slowly squeezed and heated by the accumulated rock above. Even more than agriculture, the combustion of these fossil fuels has been a spur to human population growth – but only within the past few generations.

Carbon dioxide is an important by-product of the combustion of fossil fuels, along with other gases such as sulphur dioxide and oxides of nitrogen. The processing of petroleum has led to the release of various exotic pollutants, ranging from lead to plastics. The results have included a sharp increase in temperature; widespread extinctions of animals and plants; acidification of the oceans to the detriment of coral reefs, and so on. The overall effect has been rather similar to that which might have occurred had a mantle plume punched its way to the surface through organic sediments.

In contrast to the mantle plumes whose various eructations brought the Permian to such an agonizing finale, the current human-induced disturbance will be extremely brief. Already, measures are being taken to reduce the emission of carbon dioxide, and to find energy sources other than fossil fuels. The human-caused carbon spike will be high, but very narrow – perhaps too narrow to be detectable in the very long term.

Humans have existed in numbers for so short a time that in, say, 250 million years, the remains of few, if any, will have been preserved. Those future prospectors with equipment of only the most refined sensitivity might – just *might* – be able to detect faint traces of unusual isotopes to say that, a short way through the Cenozoic Ice Age, *something happened*, but they might be unable to say precisely what.

Within the next few thousand years, *Homo sapiens* will have vanished. The cause will be, in part, the repayment of an extinction debt, long overdue. The patch of habitat occupied by humanity is nothing less than the entire Earth, and human beings have been making it progressively less habitable.

The main reason, though, will be a failure of population replacement. The human population is likely to peak during the present century, after which it will decline. By 2100, it will be less than it is today.[15] Although humans will do much to ameliorate the damage done to the Earth by their activities, they will not survive more than another few thousand to tens of thousands of years.

Human beings are already remarkably homogeneous, in terms of genetics, compared with our closest relatives among the apes. This is a sign of one or more genetic bottlenecks early in human history, followed by rapid expansion – a legacy of the near-extinction of humanity several times in the remote past.[16] Extinction will result from a combination of insufficient genetic variation due to events deep in prehistory; extinction debt due to present-day habitat loss; reproductive failure due to changes in human behaviour, and in the environment; and the more local problems faced by small groups that find themselves cut off from others of the same kind.

The glaciers will, nonetheless, grind onwards and recede, advance and recede, many more times. The human-induced injection of carbon dioxide will set back the date of the next glacial advance, but when it comes, it will be all the more sudden. Climate-induced calving of icebergs into the oceans, especially the North Atlantic, will add so much fresh water to the ocean that the Gulf Stream will seize up, and Europe and North America will be plunged into a full-scale glaciation over the course of less than a human lifetime. But no humans will be there to feel the cold.

Humans will die out some time before all the carbon dioxide generated by their frenetic activity finally seeps away. The residual greenhouse effect will take the chill off this cold snap for a time, only for it to whiplash back, the first of a rapid switchback of sudden glacial episodes and warmer spells, until, finally, the excess carbon dioxide has been soaked up and the Great Cenozoic Ice Age can proceed without further interruption.[17]

In about thirty million years, Antarctica will have drifted so far northwards that warm, equatorial water will sweep away the last remains of the ice cap. What will have been the cost, in life, of this long cold spell?

All land mammals larger than a badger will have become extinct. There will be no more large ungulates, elephants, rhinos, lions, tigers, giraffes or bears. Marsupials will be almost gone. The platypus and echidna – the egg-laying mammals that can trace their lineage deep into the Triassic – will have laid their last eggs. There will be no more primates. *Homo sapiens*, last of its kind, will have long gone.

There will be a few small birds, and quite a few lizards and snakes. Larger reptiles such as turtles and alligators will have died out, as will all remaining amphibians.

There will still be plenty of rodents, but we might have trouble recognizing them as such. Guilds of new, grazing herbivores will be able to trace their ancestry back to mice and rats. Amongst the traditional carnivores, only the smaller forms will remain, of the mongoose or ferret kind. Larger carnivores will be rodents, redux. Except, of course, for the most terrifying predators, which will evolve from giant flightless bats.[18]

There will still be fish in the sea. Sharks will cruise on much as they have done since the Devonian. There will be reefs, made of a new kind of coral or sponge.

And there will still be whales, for a while.

On the most panoramic of scales, the tale of life on Earth, with all its drama, all its comings and goings, is governed by just two things. One of them is a slow decline in the amount of carbon dioxide in the atmosphere. The other is the steady increase in the brightness of the Sun.[19]

Most life is based on the ability of photosynthetic plants to convert carbon dioxide in the atmosphere into living matter. To do this, most plants require a concentration of carbon dioxide in the atmosphere of about 150 parts per million (ppm). This is predicated on the assumption that plants convert carbon dioxide into sugars using just one kind of photosynthesis, called the 'C3' pathway. There is, however, another kind of photosynthesis, the 'C4' pathway, that can get by on much less – only 10 ppm. The problem with the C4 pathway is that it requires more energy to drive it, which is why, for the most part, plants tend to prefer the C3 pathway.[20]

A change happened a few million years ago, with the evolution of grass, especially in tropical savannah, which tends to use the more energy-profligate but carbon-dioxide-thrifty C4 pathway. Overall, and despite the occasional peak or trough, carbon dioxide had been declining steadily throughout the history of the Earth, and there came a point, midway through the Cenozoic era, when it had become so low that natural selection began to favour this otherwise unusual form of photosynthesis, despite the extra cost.

Looking further back, this is just one example of life's reaction to the challenges thrown at it by the changing conditions of the Earth to which it is confined. Behind many of these challenges lurks the steady rise in the amount of heat that reaches the Earth from the Sun; and the ups and downs – but, mainly, the downs – of carbon dioxide.

Why is carbon dioxide becoming so scarce, so precious? The answer can be framed in a single word – weathering. New rocks, thrust up through the earth to become mountains, are swiftly eroded. This process sucks carbon dioxide out of the atmosphere. The eroded rocks eventually get ground to dust that wanders its way, eventually, to the sea, where it is buried on the sea floor.

In the earliest days of the Earth, the entire surface of the planet was covered with ocean. There was little or no land to erode. Over time, however, the proportion of land has steadily increased, and, with it, the potential for weathering. Slowly, steadily, the amount of carbon dioxide removed from the atmosphere has increased, relative to the rate of its replenishment through, say, volcanic eruptions.[21]

One of life's first challenges occurred during the Great Oxidation Event, between 2.4 and 2.1 billion years ago. A spike in tectonic activity led to a sharp increase in the burial of carbon. Carbon dioxide was scrubbed from the air. The world, no longer benefiting from the greenhouse effect, was tipped into an ice age that lasted 300 million years, in which the entire world was covered in ice, from pole to pole – its first and greatest Snowball Earth episode. The severity was exacerbated by the fact that the Sun produced much less heat than it does nowadays, a fact that was to influence the future course of life on the planet.

Life responded by an increase in complexity. Individual bacteria, living in loose associations, pooled their resources, each individual concentrating on the one aspect of life it did best. It was a classic example of the division of labour, straight out of Adam Smith and *The Wealth of Nations*. Factories in which each worker concentrates on a specific task, rather than each one trying to do everything on their own, are much more efficient than the sum of their parts.

In the same way, the new nucleated, or eukaryotic cells, could achieve more with less.

Life's next major challenge came around 825 million years ago, with the break-up of the supercontinent Rodinia. As before, this led to massively increased weathering, carbon burial and another protracted series of ice ages. These ice ages caused Snowball Earth episodes, although they did not last as long as that which had frozen the planet during the Great Oxidation Event. Although there was more land to erode, the Sun was that much hotter.[22]

At that time, eukaryotes had been experimenting with a further increase in complexity, in which different eukaryotic cells would club together to make an organism composed of many different cells, each one concentrating on a different task, such as digestion, or reproduction, or defence. The evolution of animals was a direct consequence of the ice ages that followed the break-up of Rodinia.

Once again, life had responded to major environmental disruption by a thorough revision of its domestic economics. Multicellularity allowed organisms to become larger, move faster, move further and exploit more resources in a way that individual eukaryotic cells never could.

It wasn't as if eukaryotes looked at their calendars, and, seeing that it was 825 million years ago, unanimously decided to become multicellular. Multicellular creatures had evolved long before, and unicellular eukaryotes – and bacteria – would all remain extremely common. It's just that the multicellular state became more common, rather than just an oddity. A billion years ago, one would have seen the occasional frond of seaweed amid a sea of slime. By 800 million years ago, the seaweed was everywhere. By 500 million years ago, the seaweed would be jumping with animals, some large enough to see with the naked eye.

In a similar way, life is preparing for the next step change in complex evolution. As bacteria combined to create eukaryotes; as these combined to create multicellular animals, plants and fungi, so these organisms will combine to produce, in the last ages of life on Earth, a whole new kind of organism of a power and efficiency we can hardly imagine.

The seeds were sown long ago.

Not long after plants first made landfall, they found that life was much easier when they formed close associations with underground fungi called mycorrhizae, which would attach themselves to the roots of the plants. Plants would supply the fungi with nutrients from photosynthesis. The fungi would dig deep into the ground for trace minerals in exchange.[23]

Today, most land plants have associations with mycorrhizae, and, indeed, could not survive without them. When you next walk in the woods, consider that in the ground beneath your feet, the mycorrhizae of different plants have linked up to exchange nutrients, forming a wood-wide web that regulates the growth of the entire forest. The forest – with all its trees and mycorrhizae – is a single superorganism.[24]

Fungi have the potential to regulate life over very large areas. One of the largest known organisms is a specimen of the fungus *Armillaria bulbosa* whose microscopic threads have spread over an area of 15 hectares in a forest in northern Michigan. Although one would hardly be aware that it even exists, it has a total mass of more than 10,000 kilograms, and has lived for more than 1,500 years.[25] Defining this fungus as an individual, though, is hard. The threads of fungi spread, invisible, invasive, unsuspected, into every nook and corner, forming gigantic unions in secret, buried in the darkness and the soil.

Much later, as the age of dinosaurs was nearing its apogee, the world of plants underwent a quiet revolution. This was the evolution of flowers.

Flowering plants started as small creeping things in the water margins of the world, but soon became much more common. A hundred million years later, they are the dominant form of land plant.

One of the advantages of flowers is that they attract pollinators, rather than relying for their fertilization on wind and weather and chance. In flowering plants, as in so many things, life had short-circuited the environment, warping the odds in its favour.

It was probably no coincidence, therefore, that the evolution of flowers occurred at the same time as a dramatic increase in pollinating insects, especially ants, bees and wasps (collectively, the Hymenoptera) and butterflies and moths (the Lepidoptera).[26] These insects had already been in existence for millions of years, but the evolution of flowering plants turbocharged their evolution.

Some plants and their pollinators have such close associations that they cannot survive without the other. Figs, for example, cannot reproduce without their attendant fig wasps, which have built their lives around the plant. What we think of as the fruits of the fig are actually habitats created by and for the wasps.[27] There is a similarly close relationship between the yucca and its attendant moths.[28] In some respects, figs and fig wasps together form a single organism, an indissoluble union; the same can be said for the yucca and the yucca moth.

Many ants, bees and wasps have been evolving into a new and more integrated state, entirely separately from their associations with plants – for all that the evolution of flowering plants gave their evolution a boost. Many of these insects congregate into gigantic colonies in which individuals are specialized for specific tasks, such as guarding, or foraging. Significantly, reproduction is vested in a single individual, the queen. Just as in a multicellular organism, the business of reproduction is concentrated in a distinct population of cells.

Such colonies are superorganisms, and even show distinct behaviours that would otherwise be characteristics of single animals. For example, some colonies of the harvester ant *Pogonomyrmex barbatus* tend to send out foragers less during droughts than other colonies, and this restraint pays off in the founding of more daughter colonies.[29] Like humans, ants form close associations with the bacteria that live inside them, and with other animals round about. They actively cultivate gardens of fungi. They tend domesticated flocks of aphids, which they harvest for the honeydew they secrete.

Social organization is a trait linked with success.[30] The success of *Homo sapiens* might be put down to a tendency towards social organization, in which – like social insects – individuals tend to specialize in particular tasks. Such organization has the potential to accrue more resources, more easily, than would be possible for individuals acting alone. How many people, in today's world, would be able to live in any comfort were they forced to provide for even their most basic needs themselves? The same is true for the social insects. It was true before they evolved, and will be true long after humans are extinct. Indeed, the benefits of small individual size and large-scale organization will only become more important with time.

As time passes, and carbon dioxide for photosynthesis becomes scarcer, associations like this will become more common. Individual organisms will become smaller, and use resources more efficiently by forming parts of much larger social superorganisms. At the same time, plants will rely on animals to supply them with carbon dioxide, and to pollinate them. Plants with less close associations will, eventually, be starved out. Fig wasps and yucca moths are already greatly changed in their shape and behaviour from their more freewheeling, promiscuous relatives.

Plants will evolve closer associations with their

pollinators, especially if they are social insects. This change will accelerate, until the insects become simply vehicles to mediate fertilization and provide carbon dioxide. In the end, they will become little more than microscopic organs within the plant – in the same way that the mitochondria within our cells were once free-living bacteria. The reproduction of the insects will become completely synchronous with that of the plant. They will have become as one.

But the plants, too, will have changed beyond all recognition. They will, perhaps, resemble fungi, with most of their bodies as root or stem tubers underground, maybe expanded into bloated, hollow caverns in which their carbon-dioxide-producing insect partners, now more like microscopic worms, or even amoeboid-like cells, will live their entire lives, devoted to assisting the fertilization of tiny, internally produced flowers. Only occasionally will a plant send photosynthetic tissue above the ground. But with less carbon dioxide to be gathered, and withered by the increasing heat of the Sun, 'occasionally' will become 'rarely' which will become 'hardly ever'.

Some plants, though, will send tiny flowers above the ground to release and gather pollen in the wind, to maintain genetic diversity, and – perhaps – as signals, a kind of semaphore, to say that all is not yet lost.

And still the Earth moves. By 250 million years from now, the continents will once again have converged into a supercontinent, the greatest yet. Much like Pangaea, it will lie across the Equator.[31] Much of the interior will be the driest of deserts, ringed by mountain ranges of titanic height and extent.

It will show few signs of life. In the sea, life will be simpler, and a lot of it will be concentrated in the deep sea. The land will appear completely lifeless. This will be an illusion. There will still be life, but one would have to dig for it – a long, long way.

Even today, a vast legion of life lives deep underground, unregarded, deeper even than the roots of plants, deeper even than the mycorrhizae and fungi such as *Armillaria*, though they may sense it.

Deep underground live bacteria that mine minerals, eking out a very meagre existence from the energy obtained by converting them from one form to another.[32] In between the cracks, these bacteria are preyed on by a range of tiny creatures.[33] Most are roundworms, the most neglected and ignored form of animal life, for all that roundworms infest animals and plants so completely that one scientist has remarked that were all the life on Earth made transparent except for roundworms, one would still be able to see the ghostly forms of trees, animals, people, the ground itself.[34]

Life in the deep biosphere proceeds so lazily that it makes glaciers seem as sprightly as spring lambs by comparison. So slowly, in fact, as to be barely distinguishable from death. The bacteria grow very slowly, divide rarely, and can live for millennia. As the world warms, and carbon dioxide in the atmosphere becomes ever scarcer, life in the depths will speed up.

The heat itself will drive it, and so will the invasion, from above, of a new kind of organism, a barely imaginable composite of what were, in the distant past, creatures called fungi, plants and animals, but which are the very last holdouts of life near the surface of the planet. These superorganisms will put the slow-moving bacteria of the deep to work, offering safe harbour in return for energy and nutrients, for photosynthesis is now a thing of the past.

The fungal-like threads of the superorganisms will ramify through the Earth's crust, ever in search of more sustenance, more organisms to gather in, until, one day, late in the Earth's evening, the threads of all the superorganisms will have met and fused. At its last, perhaps, life will have gathered together into one single living entity, defiant against the dying of the light.

The Earth will continue to move, though more slowly, and as if with pain, because the planet is now very old, as though arthritic. The tectonic plates are not as well lubricated as they once were.

In the planet's youth, the great convective heat engines that drove continental drift were fuelled by a nuclear furnace; the slow, radioactive decay of elements such as uranium and thorium, forged in the final seconds of a supernova, and which fled to the planet's centre when it formed, so long ago. These elements have almost all gone.

The supercontinent that converges some 800 million years in the future will be the greatest in the planet's history. It will also be its last. For the continents, whose restless shifting has been the fuel for life, and, so often, its nemesis, will have finally ground to a halt.

There will be no life on the surface. Even deep underground, life is breathing its last. The last life in the sea, converging around hydrothermal vents, will be starved to death, as the mineral-rich 'smokers' of hydrogen and sulphur putter and die.

In a billion years or so, life on Earth, which has so adroitly turned every challenge to its existence into an opportunity to flourish, will have, finally, expired.[35]

EPILOGUE

To paraphrase what somebody once said in another context, the careers of all living things end in extinction. Even life itself will not endure. *Homo sapiens* will be no exception to this.

Not an exception, perhaps – but exceptional, nonetheless. Although most mammal species last for around a million years, and *Homo sapiens* even in its broadest sense has existed for less than half that time, humanity is an exceptional species. It could last for millions of years more – or suddenly drop dead next Tuesday.

The reason why *Homo sapiens* is exceptional is that it is the only species which, as far as is known, has become conscious of its place in the scheme of things. It has become aware of the damage it is doing to the world, and, therefore, has begun to take steps to limit it.

There is much current concern that *Homo sapiens* has precipitated what has been called the 'sixth' mass extinction, an event of similar magnitude to the 'Big Five', the extinctions at the end of the Permian, Cretaceous, Ordovician, Triassic and Devonian periods – events detectable in the geological record hundreds of millions of years later.

Whereas it is true that the 'background' rate of extinction – the meat-and-potatoes business in which species evolve and become extinct, each for its own reasons – has risen since the evolution of humans, and is especially high at present, human beings will need to carry on what they are doing for another 500 years for the current extinction rate to signify among the Big Five.[1] This is almost twice as long as the interval between the Industrial Revolution and the present day. Much damage has been done, but there is still time to prevent it becoming as bad as it might be, were humanity to do nothing. It is not the sixth extinction. At least, not yet.

Humanity has also precipitated an episode of global warming due, largely, to the sudden emission of carbon dioxide in the atmosphere. The effects of global warming are already being felt, and are causing significant disruption to human health and security, as well as to the lives of many different species.

One could say, of course, that it is in the nature of the climate to be changeable: that our planet has at whiles been a ball of magma; a world of water; clothed in jungle from pole to pole; and mantled in ice several miles thick.

To arrest climate change, therefore, might seem an exercise in colossal narcissistic hubris, as King Canute warned his courtiers who suggested that any king worth the name should be able to reverse the tide on command. It is tempting, when confronted with slogans such as

SAVE THE PLANET!

to retort

STOP PLATE TECTONICS!

or even

STOP PLATE TECTONICS – NOW!

After all, the Earth had been in existence for 4.6 billion years before *Homo sapiens* turned up, and will still be here long after *Homo sapiens* is gone.

Such a dyspeptic view would only be justified were humanity as unaware of its activities as, say, the first photosynthetic bacteria that adulterated the atmosphere with small but nevertheless lethal quantities of the deadly poison we now know as molecular oxygen.

Nonetheless, we *are* so aware, and are already taking steps to act in a more responsible way. Throughout the world, emission of fossil fuels is being phased out in favour of less polluting alternatives. For example, the third quarter of 2019 was the first such interval in which the UK generated more electricity from renewables than it did from power stations burning fossil fuels, and the trend is only likely to improve.[2] Cities are cleaner and greener.

Fifty years ago, when the population of the Earth was half the current headcount, there were serious worries that humanity would soon be unable to feed itself.[3] Fifty years later, however, the Earth supports twice as many people, who are, on the whole, healthier and living longer, and in a greater state of affluence than they once were. The debate has moved on to the damage caused by significant wealth inequality, rather than the absence of wealth.

Humans are beginning to support their lives more economically. They are doing this rapidly and with some enthusiasm. Although the per-capita consumption of energy is still increasing worldwide, it has declined in some high-income countries. In the UK and the United States, per-capita energy consumption peaked in the 1970s, staying more or less the same until the 2000s, since when it has decreased, and done so sharply: in the UK, per-capita energy use has declined by almost a quarter in the past twenty years alone.[4]

Humans are also better educated than they were. In 1970, only one in every five humans stayed in school until the age of twelve. It's now just over one in every two (51 per cent) and projected to reach 61 per cent by 2030.[5]

The human population, which once threatened to exceed all bounds of control, will peak in the present century, after which it will start to fall. By 2100, it will be less than it is now.[6]

More efficient technology and improvements in agriculture have been responsible for many of these things. But perhaps the single most important factor in the improvement of the human condition over the past century has been the reproductive, political and social empowerment of women, especially in developing countries. Now that women have increasing government over their own bodies, and more of a say in human affairs, humanity has doubled its workforce, improved its overall energetic efficiency, and cut its population growth.

There are many challenges still ahead. Yet humanity, as

life has always done, will respond – is responding – to them by the division of labour and by thus making fewer resources go a lot further.

Homo sapiens will, however, end up extinct, sooner or later.

There might be a let-out clause, although, when looked at closely, it will prove illusory. This book has been about life on Earth, and shows that conditions on Earth will, one day, become too hostile for any kind of life, no matter how resourceful. But I have not discussed how life might extend beyond the Earth.

Although it is known that some organisms can withstand exposure to space,[7] *Homo sapiens* is the first species from Earth which, as far as anyone knows, has deliberately set out into space; has established a crewed space station in orbit; and has set foot on another world, the Moon. It is therefore possible that human beings might regularly leave the Earth, and even live permanently in space, whether on planetary surfaces or artificial habitats.

At present, this seems unlikely. At the time of writing, only a handful of people have visited the Moon,[8] and nobody at all since 1972. This is not necessarily a reason for pessimism. When the earliest modern humans, living on the coast of southern Africa about 125,000 years ago, first developed cosmetics, learned to draw and use bows and arrows, the technology would flicker into life, only to be seemingly forgotten, sometimes for thousands of years, until, eventually, the technology was reacquired, and eventually became commonplace. It could be that such activities require people in sufficient numbers, and living sufficiently close together, to sustain such activities, to ensure that the crafts and skills required are maintained.

Space travel, seemingly abandoned, is springing back into life after a long dormancy, and might become routine. Improvements in technology have meant that space travel is no longer so costly that only governments can afford to indulge in it. Private corporations have become involved. The prospect of people visiting space just to look at the view is no longer a matter of science fiction. At first, the only customers will be the fabulously wealthy – but the same was once true for air travel.

It is worth noting how rapidly technology has evolved. For example, fifty years separated the first human landing on the Moon (July 1969) from the first transatlantic aeroplane flight (June 1919), undertaken by two brave pilots in a contraption that, to modern eyes, looks like a fragile arrangement of canvas, wood and lawnmower engines tied together with string.

But extinction will still be the fate of humanity, even if, one day, the species makes it to the stars. The colonies of humans will be very small, and separated by vast distances, raising the possibility that many will fail for lack of people, and genetic diversity, and those that succeed will, eventually, diverge into different species. There will be no escape that way.

What, then, will be the human legacy? When measured against the span of life on Earth – nothing. The whole of human history, so intense and so brief, all the wars, all the literature, all the princes and dictators in their palaces, all the joy, all the suffering, all the loves, and dreams, and achievements, will leave no more than a layer, millimetres thick, in some future sedimentary rock until that, too, is eroded to dust and comes to rest at the bottom of the ocean.

Somehow, though, this makes it all the *more* significant, all the more important that we seek to preserve what we have, to make our mayfly existence as comfortable as possible, for ourselves and our fellow species.

Star Maker, by Olaf Stapledon (1886–1950), is perhaps the most audacious piece of speculative fiction ever published. That very few have heard of it is perhaps a function of its forbidding immensity of scale (although the book itself is quite short). The story tells of the history of our cosmos, which (in the story) takes more than 400 billion years to unfold – and that's just one of several universes. The history of humanity occupies a mere paragraph.

In the story, the protagonist walks out of his cottage following an argument with his spouse. Sitting on a hillside, he is seized by a vision in which he is transported into the cosmos. Encountering other wanderers, he becomes part of a community of souls that engages in many adventures until, now accumulated as a cosmic mind, it encounters the Creator. Our universe is just an essay in the craft – other toy universes scatter the Creator's workshop. Further, greater universes are yet to come.

Returning home, the protagonist reflects on his travels. It is worth remembering that Stapledon was a confirmed pacifist who had nevertheless witnessed the horrors of war first-hand, having worked with the Friends' Ambulance

Service on the Western Front. *Star Maker* was published in 1937, when the world was sliding into another global conflict: something the protagonist discusses in the book's prologue and afterword.

How, asks the narrator, can an ordinary person face up to such inhuman horror?

'Two lights for guidance,' he offers. The first, 'our little glowing atom of community'. The second, seemingly antithetical, 'the cold light of the stars', in which matters such as world wars are of negligible account. He concludes:

> Strange, that it seems more, not less, urgent to play some part in this struggle, this brief effort of animalcules striving to win for their race some increase of lucidity before the ultimate darkness.

Therefore, do not despair. The Earth abides, and life is living yet.

Further Reading

As you'll see, this book comes with extensive notes that detail some of the primary research on which it is based. Research papers are in their nature intended to be read by other scientists. Here, in contrast, I offer some suggestions for further reading that are, hopefully, more accessible.

Benton, Michael J., *When Life Nearly Died* (London: Thames & Hudson, 2003). The story of the end-Permian extinction, in terrifying (and thus engaging) detail, with an analysis of possible causes.

Berreby, David, *Us and Them* (New York: Little, Brown, 2005). On human behaviour, in particular how easily we form mutually hostile groups and alliances. This is the finest anthropology book I have ever read. You may quote me.

Brannen, Peter, *The Ends Of The World* (London, Oneworld, 2017). The story of the various mass extinctions in Earth history.

Brusatte, Steve, *The Rise and Fall of the Dinosaurs* (London: Macmillan, 2018). A concise, current and exciting book on the very latest in dinosaur research.

Clack, Jennifer, *Gaining Ground* (Bloomington: University of Indiana Press, 2012). The guide to the origin of land vertebrates from fishy beginnings.

Dixon, Dougal, *After Man* (London: Granada, 1981). Sportive look at what wildlife might be like, 50 million years hence, if human beings disappeared today.

Fortey, Richard, *The Earth, an Intimate History* (London: HarperCollins, 2005). The entire history of our planet, from a geological perspective.

Fraser, Nicholas, *Dawn of the Dinosaurs* (Bloomington: Indiana University Press, 2006). The history of the unjustly neglected Triassic period. Evocative illustrations by Douglas Henderson.

Gee, Henry, *In Search of Deep Time* (New York: The Free Press, 1999), published in the UK as *Deep Time* (London: Fourth Estate, 2000). A book that cautions against what the book in your hands is all about – using an incomplete fossil record to tell a story. Instead, one can use the record to outline many possible stories, some of which are much more interesting than the one you thought you knew.

Gee, Henry, *The Accidental Species* (Chicago: University of Chicago Press, 2013). Your handy guide to the study of human origins and evolution, debunking a few myths and dethroning humankind from its high estate.

Gee, Henry, *Across the Bridge* (Chicago: University of Chicago Press, 2018). A guide to the origins of the vertebrates, the group of animals to which we ourselves belong.

Gee, Henry, and Rey, Luis V., *A Field Guide to Dinosaurs* (London: Aurum, 2003). A guide for travellers to the world of dinosaurs; it is very speculative. Worth it for the amazing art by Luis Rey.

Gibbons, Ann, *The First Human* (New York: Anchor, 2006). The story of research into human origins, from a leading commentator in the field.

Lane, Nick, *The Vital Question* (London: Profile, 2005). A view on how life got started, from a writer bubbling with brio.

Lieberman, Daniel, *The Story of the Human Body* (London: Allen Lane, 2013). An account of human evolution and why our modern lifestyles are so unsuited to our heritage.

McGhee, George R., Jr, *Carboniferous Giants and Mass Extinction* (New York: Columbia University Press, 2018). Lively account of the world in the Carboniferous and Permian periods.

Nield, Ted, *Supercontinent* (London: Granta, 2007). The story of continental drift and the half-billion-year-long supercontinent cycle.

Prothero, Donald R., *The Princeton Field Guide to Prehistoric Mammals* (Princeton: Princeton University Press, 2017). If you are confused about your taeniodonts and tillodonts, pantodonts and dinocerates, this is the book you need. Lovely illustrations by Mary Persis Williams.

Shubin, Neil, *Your Inner Fish* (London: Penguin, 2009). How our fishy heritage can be found in humans living today.

Stringer, Chris, *The Origin Of Our Species* (London: Allen Lane, 2011). The story of how *Homo sapiens* came to be the way it is.

Stuart, Anthony J., *Vanished Giants* (Chicago: University of Chicago Press, 2021). Detailed yet accessible overview of the extinction of most large animals towards the end of the Pleistocene. Who knew that there was a species called Yesterday's Camel?

Thewissen, J. G. M. 'Hans', *The Walking Whales* (Oakland: University of California Press, 2014). The incredible story of how a group of land animals returned to the sea and became fully marine, in just 8 million years.

Ward, Peter, and Brownlee, Donald, *The Life and Death of Planet Earth* (New York: Henry Holt, 2002). A grim prognostication of the future of life on our planet.

Wilson, Edward O., *The Social Conquest of Earth* (New York: Liveright, 2012). Passionate polemic from the founder of sociobiology on how evolution has produced superorganisms that have inherited the Earth, whether ants or humans.

Acknowledgements

After *Across the Bridge* I swore I wasn't going to write another book.

'I'm not going to write another book,' I exclaimed to my colleague David Adam. At the time, David was a news reporter and leader writer at *Nature*, where we both worked. I would often interrupt David so we could chat about books. He had written two: *The Man Who Couldn't Stop*, and *The Genius Within*.

Ignoring my protestations, David suggested that I write something about all the amazing research on fossils I have had the privilege of encountering, over the years, from my desk at *Nature*.

Still protesting that I wasn't going to write another book, I wrote the book.

It was a less a book of popular science than a tell-all exposé entitled *Let's Talk About Rex: A Personal History of Life on Earth*. My agent, Jill Grinberg at Jill Grinberg Literary Management, was keen to see what I was up to, but I counselled that as it was in the nature of a no-holds-barred, warts-and-all, up-close-and-personal revelation, I should write the whole book and share it with all those mentioned by name before allowing it out of doors on its own. She agreed. And so that's what I did.

The first stirrings of unease came from my parents, who said that it was all very nice, dear, but who, apart from those mentioned, would really care? Jill suggested that I try more of a straight narrative. So began a conversation that took months of drafts, large bytes of email and several late-night telephone conversations, before the final version emerged.

David Adam deserves the first thanks, as the book was his idea, at least to begin with. If you don't like it, blame him. Though I recall that our colleague Helen Pearson helped.

Quite a few people saw parts of the book as it was developing and some even made helpful suggestions, though, of course, the mistakes are entirely mine, as are quite a lot of the fanciful speculations. I acknowledge the wise counsel of Per Erik Ahlberg, Michel Brunet, Brian Clegg, Simon Conway Morris, Victoria Herridge, Philippe Janvier, Meave Leakey, Oleg Lebedev, Dan Lieberman, Zhe-Xi Luo, Hanneke Meijer, Mark Norell, Richard 'Bert' Roberts, De-Gan Shu, Neil Shubin, Magdalena Skipper, Fred Spoor, Chris Stringer, Tony Stuart, Tim White, Xing Xu and especially Jenny Clack, who sent in comments during her last illness. This book is dedicated to her memory.

Steve Brusatte (author of *The Rise and Fall of the Dinosaurs*) provided many useful comments and gave the draft to his students, many of whom kindly offered their own feedback. So, thank you Matthew Byrne, Eilidh Campbell, Alexiane Charron, Nicole Donald, Lisa Elliott, Karen Helliesen, Rhoslyn Howroyd, Severin Hryn, Eilidh Kirk, Zoi Kynigopoulou, Panayiotis Louca, Daniel Piroska, Hans Püschel, Ruhaani Salins, Alina Sandauer, Ruby

Stevens, Struan Stevenson, Michaela Turanski, Gabija Vasiliauskaite, and one student who chose to remain anonymous.

I apologize to anyone deserving of inclusion whose name I have omitted through oversight.

Jill has represented me since the last millennium. We've been through a lot together. When Jill sold my first trade book, *In Search of Deep Time*, I flew over to New York just to take her to dinner. Never let it be said that the age of chivalry is dead. It was under Jill's guidance that what started as a scurrilous memoir turned into the book you see here before you, such that it caught the imagination of Ravindra Mirchandani at Picador and George Witte at St Martin's Press, who took on the project at a very difficult time (the Covid pandemic of 2020–21 was in full swing). I thank Ravi, George, Jill, and all their colleagues, for pushing the project along.

The book would have been impossible had I not had the good fortune to have been offered a post at the science journal *Nature* on Friday, 11 December 1987, by the late, great John Maddox, thus allowing me to have a ringside seat at the unfolding parade of discovery during perhaps the most exciting period in the history of science.

More thanks are due to my family, for their encouragement, though my most sincere thanks go to my wife Penny, whose habitual response to exclamations that I was never going to write another book is a knowing smile.

It was Penny who shut me in my study between 7 p.m. and 9 p.m. every night (Fridays and Saturdays excepted) with a cup of tea, two digestive biscuits and my faithful dog Lulu.

I'd never have done it without them.

Notes

1 A SONG OF FIRE AND ICE

1. See for example R. M. Canup and E. Asphaug, 'Origin of the Moon in a giant impact near the end of the Earth's formation', *Nature* **412**, 708–712, 2001; J. Melosh, 'A new model Moon', *Nature* **412**, 694–695, 2001.
2. This explains why the Earth and Moon have similar compositions, and also why the Moon is rather special. Compared with most satellites in the solar system, the Moon is very large, relative to its primary (the Earth, in this case). See Mastrobuono-Battisti *et al.*, 'A primordial origin for the compositional similarity between the Earth and the Moon', *Nature* **520**, 212–215, 2012.
3. To illustrate how active the Earth remains to this day, the tectonic plate on which Australia rests is pushing northwards into Indonesia, crumpling it as it goes, at a rate twice as fast as the growth of the fingernails of Professor Bert Roberts of the University of Wollongong (or so Bert tells me – fingernail growth rates may vary). This may seem small, but it adds up over time. As Australia pushes north, the result is that the northern margin of Java is being warped downwards, submerging it. If you have flown over the north coast of Java, as I have, you can see that the northernmost districts of the city of Jakarta have been abandoned to the sea, in historical times. And Bert keeps having to clip his nails.
4. As I am telling this tale more as a story than as a scientific exercise, some of the things I'll say have more evidential support than others. The circumstances of the origin of life

are perhaps the least understood of anything else I'll discuss – except perhaps for large parts of chapter 12. This is the part that comes closest to Making Stuff Up. Part of the problem is that life itself is very hard to define, a subject tackled by Carl Zimmer in his book *Life's Edge* (Random House, 2020).

5. In particular, membranes accumulate electrical charge and allow it to dissipate by doing useful work such as driving chemical reactions. This is essentially how a battery works. Then, as now, living things were powered by electricity. It is surprisingly powerful. Given that the charge difference between the insides and outsides of cells is measurable, but the distance microscopic, the potential difference can be very large, of the order of 40–80 mV (millivolts). For a lively account of the role of electrical charge in the origin of life, and much else, see Nick Lane's book *The Vital Question*.
6. Think of teenagers, their emerging understanding and conscience increasing at the expense of the order in their immediate surroundings.
7. The oldest rocks that still survive from the first days of the Earth are between 3.8 and 4 billion years old, yet tiny but very robust crystals of a mineral called zircon are known that have survived more than 4.4 billion years, weathered from even earlier rocks that have since completely eroded into nothingness. Some of these ancient zircons bear the signs – no more than the ghost of a memory of shadows glimpsed from the corner of the eye – that life had passed that way, more than 4 billion years ago. Living things have a unique chemistry largely concerned with atoms of carbon. Almost all carbon atoms come in a variety, or 'isotope', known as carbon-12. A tiny proportion of carbon atoms are of the isotope known as carbon-13, which is slightly heavier. The kinds of chemical reactions that go on in living things are so wired that they reject carbon-13, and so are enriched in carbon-12, relative to the inorganic environment – and this discrepancy can be measured. Extremely ancient rocks that contain carbon, but slightly less carbon-13 than expected relative to carbon-12, might be telling us that life was once

present, even if actual bodily remains have disappeared – in the same way that the presence of the otherwise faded Cheshire cat can be revealed by its lingering smile. It is this kind of evidence upon which rests the claim that life existed on Earth at least 4.1 billion years ago. It comes from a zircon crystal which includes a smudge of carbon graphite with a relative richness of carbon-12 suggestive that life on Earth began so long ago that its origin antedates the earliest rocks. See Wilde *et al.*, 'Evidence from detrital zircons for the existence of continental crust and oceans on the Earth 4.4 Gyr ago', *Nature* **409**, 175–178, 2001.

8. See E. Javaux, 'Challenges in evidencing the earliest traces of life', *Nature* **572**, 451–460, 2019, for a salutary reminder of the problems of interpreting very ancient fossils.
9. At the time of writing, the earliest generally agreed-upon claim for life on Earth comes from a body of rock called the Strelley Pool Chert in Australia, which preserves the remains of not one or two fossils but an entire reef ecosystem that thrived in a warm sunlit ocean some 3.43 billion years ago. See Allwood *et al.*, 'Stromatolite reef from the Early Archaean era of Australia', *Nature* **441**, 714–718, 2006. There are other claims, going back to beyond 4 billion years, but their status is controversial.
10. At least until the evolution of animals that could graze them. Today, stromatolites survive only in those rare places that animals cannot reach. One such place is Shark Bay, in Western Australia, a body of water too salty for anything but slime to survive.
11. This is odd, because the Sun wasn't as bright then as it is now, a circumstance known as the 'Faint Young Sun Paradox'. A paradox because the Earth really ought to have been an iceball. However, the early atmosphere was full of potent greenhouse gases such as methane, which kept the temperature toasty.
12. The causes of the Great Oxidation Event remain much debated. The evidence suggests an increased period of activity that brought gases to the surface from the Earth's deep interior. See Lyons *et al.*, 'The rise of oxygen in the Earth's

early ocean and atmosphere', *Nature* **506**, 307–315, 2014; Marty *et al.*, 'Geochemical evidence for high volatile fluxes from the mantle at the end of the Archaean', *Nature* **575**, 485–488, 2019; and J. Eguchi *et al.*, 'Great Oxidation and Lomagundi events linked by deep cycling and enhanced degassing of carbon', *Nature Geoscience* doi:10.1038/s41561-019-0492-6, 2019.

13. As Joni Mitchell put it: 'by the time we got to Woodstock we were half a million strong,' and as one festival-worn music journalist added, '. . . and three hundred thousand of us were looking for the lavatory'.
14. See Vreeland *et al.*, 'Isolation of a 250 million-year-old halo-tolerant bacterium from a primary salt crystal', *Nature* **407**, 897–900, 2000; J. Parkes, 'A case of bacterial immortality?', *Nature* **407**, 844–845, 2000.
15. It is possible that this tendency was spurred on by the trauma of the Great Oxidation Event.
16. Technically, 'bacteria' (singular: bacterium) and 'archaea' (singular: archaeon) are very different kinds of organism. But all are small and of the same grade of organization, so here I am using 'bacteria' as a familiar catch-all term for both sorts.
17. See Martijn *et al.*, 'Deep mitochondrial origin outside sampled alphaproteobacteria', *Nature* **557**, 101–105, 2018.
18. The fusion between different sorts of bacteria and archaea to create nucleated cells has been traced by a kind of molecular archaeology that unpicks the fusion events (M. C. Rivera and J. A. Lake, 'The Ring of Life provides evidence for a genome fusion origin of eukaryotes', *Nature* **431**, 152–155, 2004; W. Martin and T. M. Embley, 'Early evolution comes full circle', *Nature* **431**, 134–137, 2004). The identity of the archaeon that formed the nucleus was hazy, as it would also have to have had features of nucleated cells that archaea don't have, such as a miniature skeleton of protein fibres. Such archaea have now been discovered in seabed sediments (Spang *et al.*, 'Complex archaea that bridge the gap between prokaryotes and eukaryotes', *Nature* **521**, 173–179, 2015; T. M. Embley and T. A. Williams, 'Steps on the road to

eukaryotes', *Nature* **521**, 169–170, 2015; Zaremba-Niedzwiedska *et al.*, 'Asgard archaea illuminate the origin of eukaryote cellular complexity', *Nature* **541**, 353–358, 2017; J. O. McInerney and M. J. O'Connell, 'Mind the gaps in cellular evolution', *Nature* **541**, 297–299, 2017; Eme *et al.*, 'Archaea and the origin of eukaryotes', *Nature Reviews Microbiology* **15**, 711–723, 2017). After heroic effort, these cells have been cultured in the laboratory (Imachi *et al.*, 'Isolation of an archaeon at the prokaryote-eukaryote interface', *Nature* **577**, 519–525, 2020; C. Schleper and F. L. Sousa, 'Meet the relatives of our cellular ancestor', *Nature* **577**, 478–479). Curiously, the cells are very small, but extend long tendrils that embrace nearby bacteria, some of which they require to survive; a possible precursor to the formation of cells (Dey *et al.*, 'On the archaeal origins of eukaryotes and the challenges of inferring phenotype from genotype', *Trends in Cell Biology* **26**, 476–485, 2016).

19. Even today, most eukaryotes live in the confines of a single cell. Single-celled eukaryotes include the amoeba and paramecia found in any garden pond, as well as many organisms that cause diseases, such as malaria, tropical sleeping sickness and leishmaniasis. Eukaryotes with bodies consisting of many cells stuck together include animals, plants and fungi, as well as many algae such as seaweeds, although even multicellular eukaryotes spend part of their life cycle as a single cell. You, dear reader, came from a single cell.
20. 'Sex' is entirely distinct from 'gender'. At first, the participants produced sex cells of more or less equal size. 'Gender' entered the picture when one 'mating type' produced small numbers of large sex cells, which we call 'eggs'; and the other produced large numbers of very small sex cells, which we call 'sperm'. It is in the interests of sperm-producers to fertilize as many eggs as possible, but this conflicts with the interests of egg-producers, which are inclined to be much more choosy about the quality of the sperm they allow to fertilize their limited store of eggs. The war between male and female had begun.
21. Multicellular life has evolved, quite independently, many

times (see Sebé-Pedros *et al.*, 'The origin of Metazoa: a unicellular perspective', *Nature Reviews Genetics* **18**, 498–512, 2017). Apart from animals, there are the plants and their close relatives the green algae, various kinds of red and brown algae, and assorted fungi. Most eukaryotes, however, remain unicellular – as do *all* eukaryote sex cells, including human egg and sperm cells. From a certain perspective, therefore, one could view multicellularity as a support mechanism to enable more efficient provisioning of sex cells.

22. This period of Earth's history is referred to somewhat disparagingly by geologists – who, without the promise of imminent world-shaking tectonic apocalypse, generally stay in bed – as 'The Boring Billion'.
23. Protists comprise a vast range of highly diverse single-celled eukaryote organisms that used to be condemned to a dustbin group called the 'protozoa'. As well as familiar pond life such as the amoeba and the paramecium, they include creatures important to the Earth system, including dinoflagellates that cause 'red tides', foraminifera and coccolithophores, which create mineral tests for themselves of exquisite beauty; to medicine, such as malarial parasites and trypanosomes, which cause sleeping sickness; and to general curiosity and wonderment, such as the dinoflagellate *Nematodinium* that has a perfectly formed eye with a cornea-like layer, a lens and a retina (see G. S. Gavelis, 'Eye-like ocelloids are built from different endosymbiotically acquired components', *Nature* **523**, 204–207, 2015). Protists are like Jack Russell terriers: what they lack in size, they make up for in personality.
24. See Strother *et al.*, 'Earth's earliest non-marine eukaryotes', *Nature* **473**, 505–509, 2011.
25. Lichens are associations of algae and fungi so intimate that they can be recognized as distinct species. For a delightful disquisition on lichens, see Merlin Sheldrake's book *Entangled Life: How Fungi Make Our Worlds, Change Our Minds, and Shape Our Futures* (London: The Bodley Head, 2020).
26. See N. J. Butterfield, '*Bangiomorpha pubescens* n. gen. n. sp.: implications for the evolution of sex, multicellularity, and

the Mesoproterozoic/Neoproterozoic radiation of eukaryotes', *Paleobiology* **26**, 386–404, 2000.

27. See C. Loron *et al.*, 'Early fungi from the Proterozoic era in Arctic Canada', *Nature* **570**, 232–235, 2019.
28. See El Albani *et al.*, 'Large colonial organisms with coordinated growth in oxygenated environments 2.1 Gyr ago', *Nature* **466**, 100–104, 2010.
29. Plate tectonics breathes. Every few hundred million years, continents aggregate into a single supercontinental landmass, only for this to break up again when plumes of magma from deep within the Earth puncture them from beneath, separating them again. The most recent supercontinent was Pangaea, which reached its greatest extent 250 million years or so ago. Rodinia was the one before that, Columbia earlier still; and there is evidence for even earlier ones. Everything you will ever need to know about plate tectonics may be found in *Supercontinent* by my friend Ted Nield (London: Granta, 2007). Ted assures me that the book is not about pelvic-floor exercises, as some might have thought.

2 ANIMALS ASSEMBLE

1. I've drawn much of what follows from Lenton *et al.*, 'Co-evolution of eukaryotes and ocean oxygenation in the Neoproterozoic era', *Nature Geoscience* **7**, 257–265, 2014.
2. The date for the evolution of sponges is contentious. The telltale mineralized spicules that form the skeletons of sponges rarely, if ever, appear before the Cambrian, and 'molecular' fossils thought to be diagnostic of sponges might instead have been formed by protists. See Zumberge *et al.*, 'Demosponge steroid biomarker 26-methylstigmastane provides evidence for Neoproterozoic animals', *Nature Ecology & Evolution* **2**, 1709–1714, 2018; J. P. Botting and B. J. Nettersheim, 'Searching for sponge origins', *Nature Ecology & Evolution* **2**, 1685–1686, 2018; Nettersheim *et al.*, 'Putative sponge biomarkers in unicellular Rhizaria question an early rise of animals', *Nature Ecology & Evolution* **3**, 577–581, 2019.

3. See Tatzel *et al.*, 'Late Neoproterozoic seawater oxygenation by siliceous sponges', *Nature Communications* **8**, 621, 2017. One cannot help but think of Darwin's last book, *The Formation of Vegetable Mould through the Action of Worms*, published in 1881 not long before the great man died. One would have to work hard to find a book with a less catchy title, though, having said that, I did once find on the shelves of books sent to *Nature* for review a large tome called *Activated Sludge*. But I digress. *Worms* (as it is usually known among Darwin cognoscenti) shows how the action of earthworms turning over the soil can, over immense periods of time, transform a landscape. Given that this little book encapsulated the great themes of time and change that had dominated Darwin's life in a compass that could be understood by everyone, *Worms* is a perfect capstone to his genius. Being Darwin, he actually measured the effects of worms by recording how long it took a stone placed on his back lawn to subside, through the action of worms in stirring the soil beneath.
4. Technically, the term plankton refers to a part of the ocean, rather than the organisms that live in it. The plankton is the sunlit surface layer of the ocean, rich in oxygen made by photosynthetic algae, and the communities of animals that live on the algae and on one another. Many animals that live on the ocean floor as adults (including sponges) have larvae that live in the plankton.
5. See Logan *et al.*, 'Terminal Proterozoic reorganization of biogeochemical cycles', *Nature* **376**, 53–56, 1995.
6. See Brocks *et al.*, 'The rise of algae in Cryogenic oceans and the emergence of animals', *Nature* **548**, 578–581, 2017.
7. This so-called Ediacaran fauna gets its name from the range of hills in South Australia where the first fossils of that age were discovered. Since then, Ediacaran fossils have been found at scattered locations across the world, from icy Arctic Russia, windswept Newfoundland and the deserts of Namibia to the relatively tame surroundings of central England.
8. *Dickinsonia* is now believed to have been some kind of animal, though of what kind is not clear. See Bobrovskiy *et al.*,

'Ancient steroids establish the Ediacaran fossil *Dickinsonia* as one of the earliest animals', *Science* **361**, 1246–1249, 2018.

9. See Fedonkin and Waggoner, 'The Late Precambrian fossil *Kimberella* is a mollusc-like bilaterian organism', *Nature* **388**, 868–871, 1997.
10. See Mitchell *et al.*, 'Reconstructing the reproductive mode of an Ediacaran macro-organism', *Nature* **524**, 343–346, 2015.
11. Gregory Retallack has suggested that some Ediacaran animals lived on land, a claim that is, to say the least, controversial. See G. J. Retallack, 'Ediacaran life on land', *Nature* **493**, 89–92, 2013; S. Xiao and L. P. Knauth, 'Fossils come in to land', *Nature* **493**, 28–29, 2013.
12. See Chen *et al.*, 'Death march of a segmented and trilobate bilaterian elucidates early animal evolution', *Nature* **573**, 412–415, 2019.
13. The hard parts of animals are invariably made of compounds of calcium. In clams, it is calcium carbonate. In backboned animals, such as fishes and humans, it is calcium phosphate. See S. E. Peters and R. R. Gaines, 'Formation of the "Great Unconformity" as a trigger for the Cambrian Explosion', *Nature* **484**, 363–366, 2012.
14. It's been very hard to discover what kind of animals made the stacked-conical skeletons called *Cloudina*. Rare preservation of soft tissue suggests that they were made by wormlike animals with through guts: Schiffbauer *et al.*, 'Discovery of bilaterian-type through-guts in cloudinomorphs from the terminal Ediacaran Period', *Nature Communications* **11**, 205, 2020.
15. See S. Bengtson and Y. Zhao, 'Predatorial borings in Late Precambrian mineralized exoskeletons', *Science* **257**, 367–369, 1992.
16. The arthropods comprise by far the most successful animal group. It includes the insects and their marine cousins the crustaceans; millipedes and centipedes; spiders, scorpions, mites and ticks, as well as the more obscure pycnogonids (sea spiders) and xiphosura (horseshoe crabs) and a host of extinct forms such as eurypterids and, of course, trilobites.

Close cousins to the arthropods are the curious onychophores or velvet worms, nowadays humble creatures of the leaf-litter of tropical forest floors, but which once had a noble marine history; and the tardigrades, or water bears – small creatures found living among moss that are curiously endearing for all that they are virtually indestructible, being able to withstand boiling, freezing, and the vacuum of space. If anyone from Marvel or DC Comics is reading, you missed a trick by not inventing Tardigrade Man. There, you can have that one for free.

17. *Tamisiocaris*, a relative of *Anomalocaris*, appears to have been more peaceable, having evolved fringe-like brushes on its clawlike frontal appendages suitable for gathering plankton, in the manner of the baleen of a whale, or the gill rakers of a basking shark (Vinther *et al.*, 'A suspension-feeding anomalocarid from the Early Cambrian', *Nature* **507**, 496–499, 2014). Unlike many Cambrian forms, the anomalocaridids survived into the Ordovician, in which filter-feeding species grew to the immense size of two metres (Van Roy *et al.*, 'Anomalocaridid trunk limb homology revealed by a giant filter-feeder with paired flaps', *Nature* **522**, 77–80, 2015).
18. That is perhaps less true to say now than it was in the 1980s, when Stephen Jay Gould wrote *Wonderful Life*, his ode to the Burgess Shale, a book that brought this look at early ocean life into the public spotlight. Gould suggested that many of the Burgess animals had no close relatives among animals now living.
19. See Zhang *et al.*, 'New reconstruction of the *Wiwaxia* scleritome, with data from Chengjiang juveniles', *Scientific Reports* **5**, 14810, 2015.
20. See Caron *et al.*, 'A soft-bodied mollusc with radula from the Middle Cambrian Burgess Shales', *Nature* **442**, 159–163, 2006; S. Bengtson, 'A ghost with a bite', *Nature* **442**, 146–147, 2006.
21. See M. R. Smith and J.-B. Caron, 'Primitive soft-bodied cephalopods from the Cambrian', *Nature* **465**, 469–472, 2010; S. Bengtson, 'A little Kraken wakes', *Nature* **465**, 427–428, 2010.

22. See for example Ma *et al.*, 'Complex brain and optic lobes in an early Cambrian arthropod', *Nature* **490**, 258–261, 2012. This is, of course, controversial – some researchers suggest that the reconstructed nervous system of *Fuxianhuia* is more apparent than real, and results instead from bacterial haloes left behind by the decay of internal organs. See Liu *et al.*, 'Microbial decay analysis challenges interpretation of putative organ systems in Cambrian fuxianhuiids', *Proceedings of the Royal Society of London B*, **285**: 20180051. http://dx.doi.org/10.1098/rspb.2018.005.
23. For a nuanced view of the transition between Ediacaran and Cambrian, see Wood *et al.*, 'Integrated records of environmental change and evolution challenge the Cambrian Explosion', *Nature Ecology & Evolution* 3, 528–538, 2019.
24. Although one should add that many kinds of animals known today have fossil records that are either exiguous or absent altogether. Many of these will have been soft-bodied parasites. The fossil record of nematodes, or roundworms, is almost (but not quite) blank. Of fossil tapeworms there are no signs whatsoever.

3 THE BACKBONE BEGINS

1. See Han *et al.*, 'Meiofaunal deuterostomes from the basal Cambrian of Shaanxi (China)', *Nature* **542**, 228–231, 2017. Although *Saccorhytus* is real, its internal anatomy described here is entirely conjectural, and much of the earliest history of vertebrates is a matter of debate. One of the most debatable points is whether the curious animals known as vetulicolians – we'll meet these a little later – had notochords. For the full story, including all the caveats, I invite you to read my book *Across The Bridge: Understanding the Origin of the Vertebrates* (Chicago: University of Chicago Press, 2018).
2. See Shu *et al.*, 'Primitive deuterostomes from the Chengjiang Lagerstätte (Lower Cambrian, China)', *Nature* **414**, 419–424, 2001, to which I reacted in an accompanying commentary: H. Gee, 'On being vetulicolian', *Nature* **414**, 407–409, 2001.

3. I saw this beautifully realized in an animated 3-D diorama at the Natural History Museum in Shanghai, which brought the Chengjiang Biota of Cambrian southern China to life. Among many other wonders it showed a shoal of vetulicolians flitting through open water.
4. This is the interpretation favoured by Chen *et al.* ('A possible early Cambrian chordate', *Nature* **377**, 720–722, 1995; 'An early Cambrian craniate-like chordate', *Nature* **402**, 518–522, 1999), although other interpretations are possible, as is often the case for strange and ancient fossils. See for example Shu *et al.*, 'Reinterpretation of *Yunnanozoon* as the earliest known hemichordate', *Nature* **380**, 428–430, 1996.
5. See S. Conway Morris and J.-B. Caron, '*Pikaia gracilens* Walcott, a stem-group chordate from the Middle Cambrian of British Columbia', *Biological Reviews*, **87**, 480–512, 2012.
6. Shu *et al.*, 'A *Pikaia*-like chordate from the Lower Cambrian of China', *Nature* **384**, 157–158, 1996.
7. That the form of the vertebrate body was, essentially, an uneasy alliance between two very different regions – a pharynx for feeding and a tail for motion – was grasped by Alfred Sherwood Romer in a difficult and yet visionary paper, 'The vertebrate as a dual animal – somatic and visceral', *Evolutionary Biology* **6**, 121–156, 1972.
8. Chen *et al.*, 'The first tunicate from the Early Cambrian of China', *Proceedings of the National Academy of Sciences of the United States of America* **100**, 8314–8318, 2003. The tunicates remain an overlooked but very successful group of animals to this day. Some have veered away from the cycle of life described in the text. In some species, the larva becomes mature while still mobile. These, the salps, and the larvaceans, became important in the ecology of the open oceans. Larvaceans may be small, but each one creates an intricate 'house' out of mucus; these remarkably complex structures are important parts of the ocean carbon cycle. Their remote location and fragility have posed immense challenges for imaging them, something that has only recently become possible (see Katija *et al.*, 'Revealing enigmatic mucus structures in the deep sea using DeepPIV', *Nature* **583**, 78–82,

2020). Other tunicates, however, have become colonial, with hundreds or thousands of individual animals fused into a single superorganism, whether anchored to one spot, or floating in the water. The pyrosomes, for example, form huge floating trumpet-shaped colonies. Although each individual is tiny, the colony might be big enough for divers to swim around inside them. Some tunicates can reproduce without sex, by budding. Others have sex lives of intricate complexity. The life of a tunicate is one long freewheeling marine Eden.

9. Well, *nearly* all. Some tunicates have become carnivorous, a mode of life that some creatures find tempting, no matter how seemingly unsuitable. Everyone is familiar with carnivorous plants. And, just when you thought it was safe to get back into the bath, there are even carnivorous sponges (J. Vacelet and N. Boury-Esnault, 'Carnivorous sponges', *Nature* **373**, 333–335, 1995).
10. Except cats.
11. In fishes (that is, aquatic vertebrates), this is the lateral-line system. In land vertebrates (that is, tetrapods) this has been reduced to the vestibular system of the inner ear, whose movements provide us with our sense of up and down, and where we are in the environment.
12. S. Conway Morris & J.-B. Caron, 'A primitive fish from the Cambrian of North America', *Nature* **512**, 419–422, 2014.
13. Shu *et al.*, 'Lower Cambrian vertebrates from south China', *Nature* **402**, 42–46, 1999.
14. The transformation from a filter-feeding pharynx to a set of gills might seem drastic, and it is. However, it is accomplished by one vertebrate even today, and that this the larva of the lamprey. The larva, called the ammocoete, spends its life amphioxus-like, buried tail-first in sediment. Eventually it metamorphoses, and the filter-feeding pharynx is transformed into the pharynx of the adult predator. Lampreys, and their cousins the hagfish (which, as far as is known, do not have filter-feeding larval stages), are similar to the earliest fishes, in that they are entirely soft-bodied, supported by a springy notochord, and have no jaws. Their mouths are lined

with teeth made from a horn-like substance. Lampreys and hagfish are notorious predators, showing that an absence of jaws is no barrier to life as a hunter.

15. How vertebrates got so large, in terms of the mechanism that drove it, is something of a mystery. Two possible answers, which are not mutually exclusive, are as follows. The first is that sometime in the ancestry of vertebrates, the genome (the totality of the genetic material) was duplicated, and duplicated again. Although many of the duplicated genes were subsequently lost, vertebrates have more than twice as many genes as invertebrates. The second is that embryonic vertebrates have a tissue called 'neural crest'. This consists of a gang of cells that migrate from the developing central nervous system and spread around the body, transforming – as if with magic fairy dust – otherwise undistinguished parts of the body into something new. Without neural crest, vertebrates would have no skin, face, eyes or ears. The neural crest also creates a long list of other bits and pieces, from the adrenal glands to parts of the heart. It is possible that the increase in complexity engendered by neural crest drove large size (see Green *et al.*, 'Evolution of vertebrates as viewed from the crest', *Nature* **520**, 474–482, 2015). The amphioxus is notable for its absence of neural crest, although there are hints of it in tunicates (see Horie *et al.*, 'Shared evolutionary origin of vertebrate neural crest and cranial placodes', *Nature* **560**, 228–232, 2018; Abitua *et al.*, 'Identification of a rudimentary neural crest in a non-vertebrate chordate', *Nature* **492**, 104–107, 2012).

16. The largest known invertebrate is the colossal squid (*Mesonychoteuthis hamiltoni*), believed to have a mass of around 750 kilograms, comparable with a large bear. The smallest known vertebrate in terms of length is probably *Paedophryne amauensis*, a frog from New Guinea, which measures about 7.7 millimetres long, though its mass is not known. In terms of mass, the smallest mammals are the pygmy white-toothed shrew *Suncus etruscans* (less than 2.6 grams) and the bumblebee bat *Craseonycteris thonglongyai* (less than 2 grams). You'd need 375,000 bumblebee bats to balance a colossal squid.

17. For a primer on the fossil record of early vertebrates, see P. Janvier, 'Facts and fancies about early fossil chordates and vertebrates', *Nature* **520**, 483–489, 2015.
18. Well, almost. Some clam-like animals called brachiopods have shells of calcium phosphate. And even now, vertebrates do have some tissues that are hardened with calcium carbonate – these are the 'otoliths' or 'ear stones' found in the ears of fishes, and in your inner ears, where they assist with the sensation of balance.
19. Why vertebrates chose calcium phosphate rather than calcium carbonate is unknown. However, phosphate is a vital nutrient which, unlike ubiquitous carbonate, is sometimes scarce in the sea. It could be that vertebrates used calcium phosphate as a store of phosphate as well as a means of defence. Phosphate is an essential ingredient of the genetic material, DNA. Large animals with fast-running metabolisms – such as vertebrates – need greater access to phosphate than smaller, more relaxed ones, and this might have prompted the use of calcium phosphate – as a store, as well as armour.
20. See A. S. Romer, 'Eurypterid influence on vertebrate history', *Science* **78**, 114–117, 1933.
21. See Braddy *et al.*, 'Giant claw reveals the largest ever arthropod', *Biology Letters* **4**, doi/10.1098/rsbl.2007.0491, 2007. It's a sobering thought that *Jaekelopterus* had relatives that sometimes came ashore and prowled the nighted forests of that alien epoch: see M. Whyte, 'A gigantic fossil arthropod trackway', *Nature* **438**, 576, 2005.
22. See M. V. H. Wilson and M. W. Caldwell, 'New Silurian and Devonian fork-tailed "thelodonts" are jawless vertebrates with stomachs and deep bodies', *Nature* **361**, 442–444, 1993.
23. There is a rare birth defect called cyclopia in which the face has a single median eye, no nose, and the brain is not divided into left and right halves. Foetuses with this defect are almost always stillborn and, if not, do not survive more than a few hours. This distressing condition is a result of the failure of the brain to divide into two halves and the face to widen, and it could be that it is a recollection of the early stages in facial evolution.

24. Gai *et al.*, 'Fossil jawless fish from China foreshadows early jawed vertebrate anatomy', *Nature* **476**, 324–327, 2011.
25. For a handy guide to the early evolution of jawed vertebrates, see M. D. Brazeau and M. Friedman, 'The origin and early phylogenetic history of jawed vertebrates', *Nature* **520**, 490–497, 2015.
26. Jawed vertebrates, then, have two pairs of paired fins, making four fins in all, the progenitors of our arms and legs. It is not known why we have two pairs, rather than three or four or indeed any at all. The paired fins are in addition to the unpaired midline fins such as the dorsal, anal and tail fins seen in many fishes.
27. Toothless they might have been, but placoderms were no slouches in the bedroom. There is now ample fossil evidence that placoderms had internal fertilization and even possibly live birth, like some sharks today. See for example J. A. Long *et al.*, 'Copulation in antiarch placoderms and the origin of gnathostome internal fertilization', *Nature* **517**, 196–199, 2015.
28. This does not mean that evolution was running backwards: only that much of placoderm history remains to be discovered, and presumably lies, as yet undisturbed, in early Silurian rocks. The same applies to early bony fish, found in the same Silurian deposits in southern China. For details of *Entelognathus*, see M. Zhu *et al.*, 'A Silurian placoderm with osteichthyan-like marginal jaw bones', *Nature* **502**, 188–193, 2013; and M. Friedman and M. D. Brazeau, 'A jaw-dropping fossil fish', *Nature* **502**, 175–177, 2013.
29. Well, nearly all. Even such an advanced bony fish as the coelacanth retains a notochord throughout life, just as if it were a lamprey or hagfish.
30. The cartilaginous braincases of acanthodians are preserved extremely rarely. However, enough is known of the skulls of the Devonian form *Ptomacanthus* and the Permian form *Acanthodes* to show a relationship with sharks. See M. D. Brazeau, 'The braincase and jaws of a Devonian "acanthodian" and modern gnathostome origins', *Nature* **457**, 305–308, 2009; and S. P. Davis *et al.*, '*Acanthodes* and

shark-like conditions in the last common ancestor of modern gnathostomes', *Nature* **486**, 247–250, 2012).

31. Zhu *et al.*, 'The oldest articulated osteichthyan reveals mosaic gnathostome characters', *Nature* **458**, 469–474, 2009.

4 RUNNING AGROUND

1. See Strother *et al.*, 'Earth's earliest non-marine eukaryotes', *Nature* **473**, 505–509, 2011.
2. See G. Retallack, 'Ediacaran life on land', *Nature* **493**, 89–92, 2013.
3. In what is now eastern North America.
4. The trail is called *Climactichnites* – its maker, probably something like a giant slug. See P. R. Getty and J. W. Hagadorn, 'Palaeobiology of the *Climactichnites* tracemaker', *Palaeontology* **52**, 753–778, 2009.
5. For a good overview of the early history of life on land, see W. A. Shear, 'The early development of terrestrial ecosystems' (*Nature* **351**, 283–289, 1991).
6. This was the Great Ordovician Biodiversification Event, or GOBE. For a primer on this fecund period in the history of life, see T. Servais and D. A. T. Harper, 'The Great Ordovician Biodiversification Event (GOBE): definition, concept and duration', *Lethaia* **51**, 151–164, 2018.
7. See Simon *et al.*, 'Origin and diversification of endomycorrhizal fungi and coincidence with vascular land plants', *Nature* **363**, 67–69, 1993.
8. For an excellent and very detailed account of the plants of the earliest forests, see *Carboniferous Giants and Mass Extinction: The Late Paleozoic Ice Age World* by George R. McGhee, Jr (New York: Columbia University Press, 2018).
9. See Stein *et al.*, 'Giant cladoxylopsid trees resolve the enigma of the Earth's earliest forest stumps at Gilboa', *Nature* **446**, 904–907, 2007.
10. This is entirely speculative. However, given that advanced placoderms and even members of modern groups of fishes

had appeared in the Silurian, it might not be such a stretch as all that.

11. See Zhu *et al.*, 'Earliest known coelacanth skull extends the range of anatomically modern coelacanths to the Early Devonian', *Nature Communications* **3**, 772, 2012.
12. See P. L. Forey, 'Golden jubilee for the coelacanth *Latimeria chalumnae*', *Nature* **336**, 727–732, 1988.
13. See Erdmann *et al.*, 'Indonesian "king of the sea" discovered', *Nature* **395**, 335, 1998.
14. The Australian lungfish has the largest genome of any known animal, fourteen times the size of that of humans. Although similar to the genomes of tetrapods, it is full of junk accreted during its long evolutionary history. See Meyer *et al.*, 'Giant lungfish genome elucidates the conquest of the land by vertebrates', *Nature* **590**, 284–289, 2021.
15. See Daeschler *et al.*, 'A Devonian tetrapod-like fish and the evolution of the tetrapod body plan', *Nature* **440**, 757–763, 2006.
16. See Cloutier *et al.*, '*Elpistostege* and the origin of the vertebrate hand', *Nature* **579**, 549–554, 2020.
17. See Niedzwiedzki *et al.*, 'Tetrapod trackways from the early Middle Devonian period of Poland', *Nature* **463**, 43–48, 2010.
18. Or, at the very least, Ursula Andress in *Dr No*.
19. See Goedert *et al.*, 'Euryhaline ecology of early tetrapods revealed by stable isotopes', *Nature* **558**, 68–72, 2018. It seems very strange to think of the earliest tetrapods – amphibians, essentially – emerging directly from the sea, given that most of the amphibians with which we are familiar live in fresh water. However, quite a few amphibians live in episodically salty habitats such as mangrove swamps, even today: see G. R. Hopkins and E. D. Brodie, 'Occurrence of amphibians in saline habitats: a Review and Evolutionary Perspective', *Herpetological Monographs* **29**, 1–27, 2015.
20. See C. W. Stearn, 'Effect of the Frasnian-Famennian extinction event on the stromatoporoids', *Geology* **15**, 677–679, 1987.
21. See P. E. Ahlberg, 'Potential stem-tetrapod remains from the Devonian of Scat Craig, Morayshire, Scotland', *Zoological Journal of the Linnean Society of London* **122**, 99–141, 2008.

22. See Ahlberg *et al.*, '*Ventastega curonica* and the origin of tetrapod morphology', *Nature* **453**, 1199–1204, 2008.
23. See O. A. Lebedev, [The first find of a Devonian tetrapod in USSR] *Doklady Akad. Nauk. SSSR.* 278: 1407–1413, 1984 (in Russian).
24. See Beznosov *et al.*, 'Morphology of the earliest reconstructable tetrapod *Parmastega aelidae*', *Nature* **574**, 527–531, 2019; N. B. Fröbisch and F. Witzmann, 'Early tetrapods had an eye on the land', *Nature* **574**, 494–495, 2019.
25. See Ahlberg *et al.*, 'The axial skeleton of the Devonian tetrapod *Ichthyostega*', *Nature* **437**, 137–140, 2005.
26. See M. I. Coates and J. A. Clack, 'Fish-like gills and breathing in the earliest known tetrapod', *Nature* **352**, 234–236, 1991.
27. See Daeschler *et al.*, 'A Devonian Tetrapod from North America', *Science* **265**, 639–642, 1994.
28. See M. I. Coates and J. A. Clack, 'Polydactyly in the earliest known tetrapod limbs', *Nature* **347**, 66–69, 1990.
29. See Clack *et al.*, 'Phylogenetic and environmental context of a Tournaisian tetrapod fauna', *Nature Ecology & Evolution* **1**, 0002, 2016.
30. See J. A. Clack, 'A new Early Carboniferous tetrapod with a *mélange* of crown-group characters', *Nature* **394**, 66–69, 1998.
31. See T. R. Smithson, 'The earliest known reptile', *Nature* **342**, 676–678, 1989; T. R. Smithson and W. D. I. Rolfe, '*Westlothiana* gen. nov.: naming the earliest known reptile', *Scottish Journal of Geology* **26**, 137–138, 1990.

5 ARISE, AMNIOTES

1. See Yao *et al.*, 'Global microbial carbonate proliferation after the end-Devonian mass extinction: mainly controlled by demise of skeletal bioconstructors', *Scientific Reports* **6**, 39694, 2016.
2. See J. A. Clack, 'An early tetrapod from "Romer's Gap"', *Nature* **418**, 72–76, 2002.
3. See Clack *et al.*, 'Phylogenetic and environmental context of

a Tournaisian tetrapod fauna', *Nature Ecology & Evolution* **1**, 0002, 2016.

4. See Smithson *et al.*, 'Earliest Carboniferous tetrapod and arthropod faunas from Scotland populate Romer's Gap', *Proceedings of the National Academy of Science of the United States of America*, **109**, 4532–4537, 2012.
5. See Pardo *et al.*, 'Hidden morphological diversity among early tetrapods', *Nature* **546**, 642–645, 2017.
6. Very slow indeed – it might have taken several years.
7. Those insects that appear to have a single pair of wings have a second pair in a disguised form. In beetles, the front pair of wings has evolved to become tough wing-covers. In flies, the second pair of wings is reduced to a pair of tiny organs that rotate rapidly and serve as gyroscopes, accounting for their legendary manoeuvrability and explaining why they are so very hard to hit with your rolled-up newspaper.
8. See A. Ross, 'Insect Evolution: the Origin of Wings', *Current Biology* **27**, R103–R122, 2016. Palaeodictyopterans are sadly no longer with us – they died out at the end of the Permian, along with the forests that nourished them.
9. I am indebted to *Carboniferous Giants and Mass Extinction* by George McGhee, Jr (Columbia University Press, 2018) for its vivid and detailed descriptions of life in the great coal forests.
10. A dramatic window on life in the early Carboniferous, at the very beginnings of the great coal forests, comes from a limestone quarry at East Kirkton, near Edinburgh, Scotland. Some 330 million years ago it was close to the Equator, and has yielded remarkable remains of early amphibians; amniotes (and their close relatives); as well as arthropods such as millipedes, scorpions, the earliest known harvestman spider, and fragments of giant eurypterids. The treasure trove was a function of the unusual geological conditions: the area was geologically active, with hot springs – which must have been inimical to water life – and nearby, active volcanoes that would occasionally coat everything in hot ash. At the same time, there was a lot of black, oozy, oxygen-free mud in which creatures could be preserved almost intact. There were

no fishes. For the geology and an overview see Wood *et al.*, 'A terrestrial fauna from the Scottish Lower Carboniferous', *Nature* **314**, 355–356, 1985; A. R. Milner, 'Scottish window on terrestrial life in the early Carboniferous', *Nature* **314**, 320–321, 1985). Apart from the near-amniote *Westlothiana*, and many other forms, East Kirkton has produced a baphetid – a member of a group of animals that was neither amniote nor amphibian, illustrating the fact that, in those days, it was hard, just by looking at them, to work out which creature belonged in what group. And we do not know which one laid what kind of egg, or if there was any transitional form between amphibian egg and amniote egg. That creature was named, in reference to its surroundings, *Eucritta melanolimnetes* – the Creature from the Black Lagoon (J. A. Clack, 'A new early Carboniferous tetrapod with a mélange of crown-group characters', *Nature* 394, 66–69, 1998).

11. Although I have strayed into speculation here, modern amphibians have adopted all these strategies, and more, so it is reasonable to suggest that their extinct relatives did much the same.
12. We humans do not lay eggs, but we do retain the various membranes, including the amnion. This is the sac in which the foetus develops. When an expectant mother announces that her 'waters have broken', this is the amniotic sac rupturing, an event soon followed by hatching. Or, in our case, birth.
13. Even the shells of dinosaur eggs were leathery; as were the largest fossil eggs known, possibly laid by a marine reptile. See Norell *et al.*, 'The first dinosaur egg was soft', *Nature* doi.org/10.1038/s41586-020-2412-8, 2020; Legendre *et al.*, 'A giant soft-shelled egg from the Late Cretaceous of Antarctica', *Nature* doi.org/10.1038/s41586-020-2377-7, 2020; J. Lindgren and B. P. Kear, 'Hard evidence from soft fossil eggs', *Nature* doi.org/10.1038/d41586-020-01732-8, 2020.
14. For much more detail on the formation of Pangaea and its consequences, especially the collapse of almost all life at the end of the Permian, see Ted Nield's book *Supercontinent*

(London: Granta, 2007) and Michael J. Benton's *When Life Nearly Died* (London: Thames & Hudson, 2003).

15. See Sahney *et al.*, 'Rainforest Collapse triggered Carboniferous tetrapod diversification in Euramerica', *Geology* **38**, 1079–1082, 2010.
16. See M. Laurin and R. Reisz, '*Tetraceratops* is the earliest known therapsid', *Nature* **345**, 249–250, 1990.
17. Entirely distinct from 'theropsids', not to say 'therapists'.
18. Magma plumes are different from the regular bump and grind of continental drift. They arise from very deep in the Earth, where the Earth's mantle meets the core. Local temperature anomalies cause magma to rise until it meets the crust, which it melts. Several notable features of the present-day Earth are caused by magma plumes, such as the island of Iceland (where the plume happens to coincide with a mid-ocean spreading centre) and Hawaii (where the plume has surfaced in the centre of a tectonic plate). Plumes last for millions of years, but are not always active. This means that a static plume underneath a moving tectonic plate can create a chain of islands of successively older ages – like the needle in a sewing machine that creates a chain of stitches in a moving piece of fabric. For example, the Pacific Plate has been moving slowly north-westwards across the mantle plume, creating a chain of islands that are successively older, the further one moves from the plume's hot spot. This means that the Big Island of Hawaii, at the south-eastern end of the chain, sits athwart the plume and is still volcanically active; the volcanoes of the islands to the north-west, such as Maui and Oahu, are dormant or extinct, and the islands get progressively smaller and more eroded as one moves yet further north-west, to end up as no more than tiny atolls, such as Laysan and Midway, at the extremities. These last islands were once as big and spectacular as Hawaii itself, but the moving plate, having met the plume, moves on, leaving weather and time to degrade the evidence of its passing. The Big Island will slowly decay, as the plate continues to drift north-west, and the volcanic activity will become concentrated in the rising Lo'ihi seamount, some 975 metres

beneath the waves off the south-eastern shore of the Big Island.

19. This is the phenomenon known as 'coral bleaching', observable today as a consequence of the currently increased concentration of carbon dioxide in the atmosphere.
20. All modern coral reefs are made of another kind of stony coral, which evolved in the Triassic. Rugose and tabulate corals – their diversity, and the diversity they supported – are no more than fossilized memories.
21. Grasby *et al.*, 'Toxic mercury pulses into late Permian terrestrial and marine environments', *Geology* doi.org/10.1130/G47295.1, 2020.
22. Feather stars are the free-living forms of sea lilies, or crinoids, nowadays found mainly in deep water.
23. The tale of *Miocidaris*, the last genus of sea urchin, is told by Erwin, in 'The Permo-Triassic Extinction', *Nature* **367**, 231–236, 1994.

6 TRIASSIC PARK

1. Dinosaurs, which evolved towards the end of the Triassic, always get top billing in any discussion of prehistoric life. This is a shame, as the range of reptilian forms that lived in the Triassic was in all ways except brute size the equal of dinosaurs in diversity and, from our perspective, strangeness. This is reflected in the fact that books about dinosaurs are two a penny, yet works on the Triassic are much scarcer. I refer you in especial to the masterly treatise by Nicholas Fraser, illustrated by Douglas Henderson, which is now very hard to come by, and whose title *Life In The Triassic* had to be relegated to the subtitle, so that it could be billed, teasingly, as *Dawn of the Dinosaurs* (Bloomington: Indiana University Press, 2006). I got my copy second-hand. It had been deleted from the public library in Pinellas Park, Florida. I bet it still has shelves groaning with dinosaur books.
2. See Li *et al.*, 'An ancestral turtle from the Late Triassic of

southwestern China', *Nature* **456**, 497–501, 2008; Reisz and Head, 'Turtle origins out to sea', *Nature* **456**, 450–451, 2008.

3. See R. Schoch and H-D. Sues, 'A Middle Triassic stem-turtle and the evolution of the turtle body plan', *Nature* **523**, 584–587, 2015. A recent reappraisal advances the idea that *Pappochelys* was more likely to have been a burrower on land than a swimmer at sea: see Schoch *et al*., 'Microanatomy of the stem-turtle *Pappochelys rosinae* indicates a predominantly fossorial mode of life and clarifies early steps in the evolution of the shell', *Scientific Reports* **9**, 10430, 2019.
4. See Li *et al*., 'A Triassic stem turtle with an edentulous beak', *Nature* **560**, 476–479, 2018.
5. See Neenan *et al*., 'European origin of placodont marine reptiles and the evolution of crushing dentition in Placodontia', *Nature Communications* **4**, 1621, 2013.
6. If you think I am making this up, you'd only be partly right. The anatomy of drepanosaurs defies description. They have been touted as swimmers, as tree-climbers with prehensile tails; as burrowers – and, with their uncannily birdlike skulls, early relatives of birds.
7. See for example Chen *et al*., 'A small short-necked hupehsuchian from the Lower Triassic of Hubei Province, China', *PLoS ONE* **9**, e115244, 2014.
8. See E. L. Nicholls and M. Manabe, 'Giant ichthyosaurs of the Triassic – a new species of *Shonisaurus* from the Pardonet Formation (Norian: Late Triassic) of British Columbia', *Journal of Vertebrate Paleontology* **24**, 838–849, 2004.
9. See Simões *et al*., 'The origin of squamates revealed by a Middle Triassic lizard from the Italian Alps', *Nature* **557**, 706–709, 2018.
10. See Caldwell *et al*., 'The oldest known snakes from the Middle Jurassic-Lower Cretaceous provide insights on snake evolution', *Nature Communications* **6**, 5996, 2015.
11. See M. W. Caldwell and M. S. Y. Lee, 'A snake with legs from the marine Cretaceous of the Middle East', *Nature* **386**, 705–709, 1997.
12. See S. Apesteguía and H. Zaher, 'A Cretaceous terrestrial

snake with robust hindlimbs and a sacrum', *Nature* **440**, 1037–1040, 2006.

13. The common ancestor of dinosaurs and pterosaurs might have been a rather small animal, which could explain the tendency towards warm-bloodness as well as the fluffiness seen in both groups. See Kammerer *et al.*, 'A tiny ornithodiran archosaur from the Triassic of Madagascar and the role of miniaturization in dinosaur and pterosaur ancestry', *Proceedings of the National Academy of Sciences of the United States of America* doi.org/10.1073/pnas.1916631117, 2020. Discovering the roots of the pterosaur lineage in particular, however, has been a challenge. The earliest pterosaurs appear in the fossil record fully formed. However, a clue to their ancestry lies in the discovery of small, bipedal archosaurs called lagerpetids. These plainly could not have flown, but share details of their brain and wrist anatomy exclusively with pterosaurs, suggesting that lagerpetids were more closely related to pterosaurs than to other animals. See Ezcurra *et al.*, 'Enigmatic dinosaur precursors bridge the gap to the origin of Pterosauria', *Nature* **588**, 445–449, 2020; and K. Padian, 'Closest relatives found for pterosaurs, the first flying vertebrates', *Nature* **588**, 400–401, 2020.
14. It's all in a wonderful paper by C. D. Bramwell and G. R. Whitfield entitled 'Biomechanics of *Pteranodon*', originally published in 1984 in *Philosophical Transactions of the Royal Society of London B* **267**, http://doi.org/10.1098/rstb.1974.0007. When I was a student at Leeds University in the early 1980s, my professor, Robert McNeill Alexander, set me a library project on flying reptiles. Alexander was the leading expert on biomechanics – the science of animal movement – so my dissertation was full of aerodynamics: lift, drag, glide polars, slope soaring and ground effect. It was Alexander who pointed me towards Bramwell and Whitfield's classic paper.
15. Bats – the only extant mammals that fly, rather than simply glide – don't have birdlike keeled breastbones either.
16. See S. J. Nesbitt *et al.*, 'The earliest bird-line archosaurs and

the assembly of the dinosaur body plan', *Nature* **544**, 484–487, 2017.

17. The earliest silesaur was *Asilisaurus*, from the Middle Triassic of Tanzania. See Nesbitt *et al.*, 'Ecologically distinct dinosaurian sister group shows early diversification of Ornithodira', *Nature* **464**, 95–98, 2010.
18. See Sereno *et al.*, 'Primitive dinosaur skeleton from Argentina and the early evolution of Dinosauria', *Nature* **361**, 64–66, 1993.

7 DINOSAURS IN FLIGHT

1. For a detailed examination of the biomechanics involved in the transition from bipedal walking to flight, see Allen *et al.*, 'Linking the evolution of body shape and locomotor biomechanics in bird-line archosaurs', *Nature* **497**, 104–107, 2013.
2. See J. F. Bonaparte and R. A. Coria, 'Un nuevo y gigantesco sauropodo titanosaurio de la Formacion Rio Limay (Albiano-Cenomaniano) de la Provincio del Neuquen, Argentina', *Ameghiniana* **30**, 271–282, 1993.
3. See R. A. Coria and L. Salgado, 'A new giant carnivorous dinosaur from the Cretaceous of Patagonia', *Nature* **377**, 224–226, 1995.
4. To move at anything more than a slow amble, *T. rex* would have needed unfeasibly large hindlimbs – its leg extensor muscles would have had to have had 99 per cent the mass of the entire animal – and that figure is for *each leg*, not both. See J. R. Hutchinson and M. Garcia, '*Tyrannosaurus* was not a fast runner', *Nature* **415**, 1018–1021, 2002.
5. See Erickson *et al.*, 'Bite-force estimation for *Tyrannosaurus rex* from tooth-marked bones', *Nature* **382**, 706–708, 1996; P. M. Gignac and G. M. Erickson, 'The biomechanics behind extreme osteophagy in *Tyrannosaurus rex*', *Scientific Reports* **7**, 2012, 2017.
6. Fossilized faeces, or coprolites, of giant carnivorous dinosaurs, most likely *Tyrannosaurus rex*, have been found. One measures 44 centimetres long by 13 centimetres wide and

16 centimetres deep, and has a mass of more than 7 kilograms, up to half of which consists of bone fragments. See Chin *et al.*, 'A king-sized theropod coprolite', *Nature* **393**, 680–682, 1998.

7. See Schachner *et al.*, 'Unidirectional pulmonary airflow patterns in the savannah monitor lizard', *Nature* **506**, 367–370, 2014.
8. See for example P. O'Connor and L. Claessens, 'Basic avian pulmonary design and flow-through ventilation in non-avian theropod dinosaurs', *Nature* **436**, 253–256, 2005, which reports how air sacs penetrated the long bones of *Majungatholus atopus*, a carnivorous dinosaur that lived in what is now Madagascar.
9. Imagine a sugar cube that measures a centimetre on each side. Its volume will be 1 × 1 × 1 = 1 cubic centimetre. A cube has six sides of equal area, so the surface area of our sugar cube will be 6 × 1 × 1 = 6 square centimetres, a ratio of 6:1. Now, imagine a sugar cube that measures two centimetres on each side. The volume has grown to 2 × 2 × 2 = 8 cubic centimetres, but the surface area will be 6 × 2 × 2 = 24 square centimetres, a ratio of 24:8, or 3:1. In brief, by doubling the unit size of the cube, the surface area has halved relative to the volume.
10. Consider: the total surface area of a human being on the outside is between 1.5 and 2 square metres, but the surface area of pair of human lungs is between 50 and 75 square metres.
11. This phenomenon, known as gigantothermy, has been used to explain how large and ostensibly cold-blooded animals such as leatherback turtles – which may have a mass of more than 900 kilograms – can keep warm even when swimming through cold seas. See Paladino *et al.*, 'Metabolism of leatherback turtles, gigantothermy, and thermoregulation of dinosaurs', *Nature* **344**, 858–860, 1990.
12. For a very insightful discussion on this subject see Sander *et al.*, 'Biology of the sauropod dinosaurs: the evolution of gigantism', *Biological Reviews of the Cambridge Philosophical Society* **86**, 117–155, 2011.

13. The furry pelage of pterosaurs might in fact be a variety of feathery plumage, too: see Yang *et al.*, 'Pterosaur integumentary structures with complex feather-like branching', *Nature Ecology & Evolution* **3**, 24–30, 2019.
14. If not feathers, then hair, or, if living a streamlined life in the sea, blubber. Sea mammals such as whales and seals have a thick coat of blubber which both insulates the core and presents an aerodynamic shape, smoothing out any lumps and bumps. The extinct marine reptiles known as ichthyosaurs, which looked very like modern dolphins, are now known to have had coats of blubber, presumably for the same reasons. See Lindgren *et al.*, 'Soft-tissue evidence for homeothermy and crypsis in a Jurassic ichthyosaur', *Nature* **564**, 359–365, 2018.
15. See Zhang *et al.*, 'Fossilized melanosomes and the colour of Cretaceous dinosaurs and birds', *Nature* **463**, 1075–1078, 2010; Xu *et al.*, 'Exceptional dinosaur fossils show ontogenetic development of early feathers', *Nature* **464**, 1338–1341, 2010; Li *et al.*, 'Melanosome evolution indicates a key physiological shift within feathered dinosaurs', *Nature* **507**, 350–353, 2014; Hu *et al.*, 'A bony-crested Jurassic dinosaur with evidence of iridescent plumage highlights complexity in early paravian evolution', *Nature Communications* **9**, 217, 2018.
16. Matters are different in the sea, where water allows for the support of much larger bodies than is possible on land, and live-bearing is favoured, because returning ashore to lay eggs, as turtles do, is extremely risky. This may explain why the earliest jawed vertebrates – the placoderms – were live-bearers, and the habit is seen in many fishes, such as sharks. Ichthyosaurs, the amniotes that returned to the sea in the Triassic and became very like whales, bore live young. Whales themselves, are, of course, live-bearers, like nearly all mammals, and evolved to become the largest animals known, eclipsing even the largest dinosaurs.
17. *Kayentatherium*, from the early Jurassic of Arizona, was a tritylodont – a member of a late therapsid group that was very close to being a mammal without actually making the

grade. Although very likely to have been furry, it almost certainly laid eggs. A single *Kayentatherium* litter could contain at least thirty-eight individuals – far greater than any mammalian litter. See Hoffman and Rowe, 'Jurassic stem-mammal preinates and the origin of mammalian reproduction and growth', *Nature* **561**, 104–108, 2018.

18. See Schweitzer *et al.*, 'Gender-specific reproductive tissue in ratites and *Tyrannosaurus rex*', *Science* **308**, 1456–1460, 2005; Schweitzer *et al.*, 'Chemistry supports the identification of gender-specific reproductive tissue in *Tyrannosaurus rex*', *Scientific Reports* **6**, 23099, 2016.
19. See G. E. Erickson *et al.*, 'Gigantism and comparative life history parameters of tyrannosaurid dinosaurs', *Nature* **430**, 772–775, 2004.
20. Live-bearing would be a severe hindrance to flight in birds. It is perhaps no coincidence that pterosaurs – the flying cousins of dinosaurs – also laid eggs (see Ji *et al.*, 'Pterosaur egg with a leathery shell', *Nature* **432**, 572, 2004), as well as evolving feather-like insulation and a very lightweight airframe.
21. Waterfowl such as swans and geese take off like this, and one can see from their efforts that birds only a little larger would not be able to become airborne in this way. This is how aeroplanes do it, too, even without flapping, and is why large airliners have enormous engines capable of incredible thrust. It takes a lot of energy to get a jumbo jet airborne. Of course, we both know, whenever we see an airliner in flight, that no amount of physics could get such a vast structure into the air. Airliners fly only because we believe they can. If we stopped believing, they'd plummet out of the sky. That's what I really think. But don't tell anybody. It's our little secret, OK?
22. Tim White reminds me that some wingless ants, although very small and which might otherwise be counted as aimless aeroplankton, can glide, after a fashion. See Yanoviak *et al.*, 'Aerial manoeuvrability in wingless gliding ants (*Cephalotes atratus*)', *Proceedings of the Royal Society of London B*, **277**, 2010, https://doi.org/10.1098/rspb.2010.0170.

23. See for example Meng *et al.*, 'A Mesozoic gliding mammal from northeastern China', *Nature* **444**, 889–893, 2006.
24. The smallest parachutists, though, use threads and bristles instead of continuous wing-like sheets. One thinks of spiders, using long threads to carry them through the air: or the bristled seed that lovelorn youths since time immemorial have blown from dandelion clocks. Each dandelion seed can be carried for miles with a stalk ending in a tuft rather like a chimney sweep's brush. Rather than trying to trap all the air beneath it, the tuft allows most of it through, and here is where the magic happens. The airflow let through the tuft becomes turbulent, forming a kind of smoke ring above the tuft. This ring, the shape of a doughnut squeezed from side to side, is an area of low pressure, a cyclone in miniature, a storm centre writ small. It literally sucks the tuft upwards, slowing its rate of descent. See Cummins *et al.*, 'A separated vortex ring underlies the flight of the dandelion', *Nature* **562**, 414–418, 2018.
25. The earliest stages in ancient parachuting have been studied in modern cats in that most contemporary of wildlife habitats – Manhattan. Veterinarians in New York are familiar with a pattern of feline injuries known as 'high-rise syndrome' sustained by venturesome cats that fall out of high windows. New York vets have plotted the severity of feline injuries against the height from which they had fallen in each case. Injuries tend to get more severe as one goes from the ground upwards, but there gets to be a point above which the cats' injuries get less severe, not more. The vets cite a case of a cat that fell thirty-two storeys and walked away with no more than light injury to its chest, a tooth and its dignity. It's not for nothing that cats have a proverbial nine lives. What seems to happen is that as a cat falls, its muscles relax and its paws splay out sideways, forming a kind of parachute. The cat may sustain injuries to its jaw and thorax, but may yet live. See W. O. Whitney and C. J. Mehlhaff, 'High-rise syndrome in cats', *Journal of the American Veterinary Medical Association* **192**, p. 542, 1988.
26. See F. E. Novas and P. F. Puertat, 'New evidence concerning

avian origins from the Late Cretaceous of Patagonia', *Nature* **387**, 390–392, 1997.

27. See Norell *et al.*, 'A nesting dinosaur', *Nature* **378**, 774–776, 1995.
28. See for example Xu *et al.*, 'A therizinosauroid dinosaur with integumentary structures from China', *Nature* **399**, 350–354, 1999, which describes featherlike structures in *Beipaiosaurus*, one of the very strange therizinosaurs. These were weird, ungainly theropods that had become herbivores, and would have been about as aerodynamic as a breeze block. See also Xu *et al.*, 'A gigantic bird-like dinosaur from the Late Cretaceous of China', *Nature* **447**, 844–847, 2007, on *Gigantoraptor*, an 8-metre, 1,400-kilogram monster that belonged to the otherwise lithe and birdlike oviraptorosaurids. This creature would certainly have been flightless – whether it had feathers is unknown.
29. Ken Dial of the University of Montana explored how the chicks of a kind of partridge called the chukar use their wings to help them run up very steep slopes, a kind of locomotion called 'wing-assisted incline running' – which would have been useful to a small, defenceless animal for escaping predators. See Dial *et al.*, 'A fundamental avian wing-stroke provides a new perspective on the evolution of flight', *Nature* **451**, 985–989, 2008.
30. Xu *et al.*, 'The smallest known non-avian theropod dinosaur', *Nature* **408**, 705–708, 2000; Dyke *et al.*, 'Aerodynamic performance of the feathered dinosaur *Microraptor* and the evolution of feathered flight', *Nature Communications* **4**, 2489, 2013.
31. Hu *et al.*, 'A pre-*Archaeopteryx* troödontid theropod from China with long feathers on the metatarsus', *Nature* **461**, 640–643, 2009.
32. See F. Zhang *et al.*, 'A bizarre Jurassic maniraptoran from China with elongate, ribbon-like feathers', *Nature* **455**, 1105–1108, 2008.
33. See Xu *et al.*, 'A bizarre Jurassic maniraptoran theropod with preserved evidence of membranous wings', *Nature* **521**, 70–73, 2015; and Wang *et al.*, 'A new Jurassic

scansoriopterygid and the loss of membranous wings in theropod dinosaurs', *Nature* **569**, 256–259, 2019.

34. To be sure, there are no known secondarily flightless bats, although the mystacinid bats of New Zealand live most of the time on the ground. Unless one counts the possible reconstructions of some giant pterosaurs as flightless, there were no known secondarily flightless pterosaurs, either.
35. See Field *et al.*, 'Complete *Ichthyornis* skull illuminates mosaic assembly of the avian head', *Nature* **557**, 96–100, 2018.
36. See Altangerel *et al.*, 'Flightless bird from the Cretaceous of Mongolia', *Nature* **362**, 623–626, 1993, for the discovery of the first one of these oddities, *Mononykus*; and Chiappe *et al.*, 'The skull of a relative of the stem-group bird *Mononykus*', *Nature* **392**, 275–278, 1998, for the discovery of another, *Shuvuuia*, to show that the first wasn't a fluke.
37. See Field *et al.*, 'Late Cretaceous neornithine from Europe illuminates the origins of crown birds', *Nature* **579**, 397–401, 2020, and the accompanying commentary by K. Padian, 'Poultry through time', *Nature* **579**, 351–352, 2020. Another Cretaceous bird that may be an early representative of the waterfowl is *Vegavis*, from Antarctica: see Clarke *et al.*, 'Definitive fossil evidence for the extant avian radiation in the Cretaceous', *Nature* **433**, 305–308, 2005. *Vegavis* had a well-developed syrinx (Clarke *et al.*, 'Fossil evidence of the avian vocal organ from the Mesozoic', *Nature* **538**, 502–505, 2016; P. M. O'Connor, 'Ancient avian aria from Antarctica', *Nature* **538**, 468–469, 2016), the distinctive vocal organ of birds that produces everything from the honk of a goose to the trill of nightingales which, legend has it, can be heard in Berkeley Square, but only when angels dine at the Ritz.
38. Note the 'almost', for biology treasures its exceptions. There is at least one record of a ceratopsian dinosaur from Europe. See, for example, Ösi *et al.*, 'A Late Cretaceous ceratopsian dinosaur from Europe with Asian affinities', *Nature* **465**, 466–468, 2010; Xu, 'Horned dinosaurs venture abroad', *Nature* **465**, 431–432, 2010.
39. See Sander *et al.*, 'Bone histology indicates insular dwarfism

in a new Late Jurassic sauropod dinosaur', *Nature* **441**, 739–741, 2006.

40. See Buckley *et al.*, 'A pug-nosed crocodyliform from the Late Cretaceous of Madagascar', *Nature* **405**, 941–944, 2000.
41. See M. W. Frohlich and M. W. Chase, 'After a dozen years of progress the origin of angiosperms is still a great mystery', *Nature* **450**, 1184–1189, 2007.
42. See for example Rosenstiel *et al.*, 'Sex-specific volatile compounds influence microarthropod-mediated fertilization of moss', *Nature* **489**, 431–433, 2012.
43. One thinks of Io and Europa, both moons of Jupiter, but otherwise quite different. Io's surface is constantly resurfaced by volcanic activity; Europa's, by ice seeping from a subsurface ocean.
44. See Bottke *et al.*, 'An asteroid breakup 160 Myr ago as the probable source of the K/T impactor', *Nature* **449**, 48–53, 2007; P. Claeys and S. Goderis, 'Lethal billiards', *Nature* **449**, 30–31, 2007.
45. See Collins *et al.*, 'A steeply inclined trajectory for the Chicxulub impact', *Nature Communications* **11**, 1480, 2020.
46. The last ichthyosaurs expired some millions of years earlier, thus avoiding all the apocalyptic fuss and brouhaha.
47. See Lowery *et al.*, 'Rapid recovery of life at ground zero of the end-Cretaceous mass extinction', *Nature* **558**, 288–291, 2018.

8 THOSE MAGNIFICENT MAMMALS

1. See J. A. Clack, 'Discovery of the earliest-known tetrapod stapes', *Nature* **342**, 425–427, 1989; A. L. Panchen, 'Ears and vertebrate evolution', *Nature* **342**, 342–343, 1989; J. A. Clack, 'Earliest known tetrapod braincase and the evolution of the stapes and fenestra ovalis', *Nature* **369**, 392–394, 1994. The middle ear of *Acanthostega*'s relative, *Ichthyostega*, appears to have been modified into a kind of aquatic hearing organ unlike anything else seen in evolution (Clack *et al.*, 'A uniquely specialized ear in a very early tetrapod', *Nature* **425**, 65–69, 2003).

2. Whereas the spiracle had conducted water in and out, communicating between the outside world and the mouth cavity, the eardrum formed a barrier, defining the outer limits of the middle ear. The middle ear did, however, retain a connection with the mouth cavity. You can feel it whenever you swallow: the action equalizes the pressure between the middle ear and the outside world, by way of a connection called the Eustachian tube. This is why sound is muzzy when you have a head cold. The Eustachian tube fills with mucus, making it difficult to equalize the pressure, so the eardrum works less efficiently. It also explains why ascending and descending in aircraft can be so painful. Even in a pressurized cabin, sudden changes in atmospheric pressure are sufficient to put the eardrum under tension, which is why it's a good idea to swallow, pushing air through the Eustachian tube and clearing any blockages. Blowing the nose has the same effect. In adult humans, the Eustachian tube is angled downwards from the middle ear to the back of the throat, so that mucus naturally drains out. In small children, however, the Eustachian tube is more or less horizontal. Small children being as they are – adorable snot-nosed vectors of contagion – mucus gets trapped in the Eustachian tube, leading to a phenomenon known as 'glue ear', which can be treated by making small holes in the eardrum. These heal up, by which time the child will have grown out of the problem.
3. The male white bellbird (*Procnias albus*) of the Brazilian Amazon makes the loudest noises of any perching bird, and does so when right up close to the female it intends to woo. The hapless inamorata experiences a sound pressure of 125 decibels. (J. Podos and M. Cohn-Haft, 'Extremely loud mating songs at close range in white bellbirds', *Current Biology* doi.org/10.1016/j.cub.2019.09.028, 2019). In humans, this is loud enough to be painful. The *Guinness Book of Records* reported sound-pressure levels of 117 dB during a concert of my favourite band, Deep Purple, at the Rainbow Theatre in London in 1972, during which three members of the audience passed out. The record has reportedly since been broken, though as the *Guinness Book of Records* no longer

reports such feats, most subsequent reports (such as the 136 dB at a Kiss Koncert in Ottawa in 2009) are unofficial. However, given that decibels increase logarithmically, the call of the bellbird is nearly three times as loud as Deep Purple's ear-shattering performance. One wonders why the female puts up with all that racket.

4. For reference, the 'A' above middle C on the piano is conventionally tuned to a frequency of 440 hertz (Hz). Frequency doubles with each octave, so the A an octave above is 880 Hz; two octaves, 1760 Hz (or 1.76 kilohertz, kHz); three octaves, 3520 Hz (3.52 kHz). After that an ordinary piano keyboard runs out of notes. If there were another A, it would be 7040 Hz (7.04 kHz), which is above the highest notes that most birds can usually hear. Human children can hear pitches up to 20 kHz, though pitch sensitivity declines in adulthood. Especially in those of us who spent their youth listening to Deep Purple.
5. The rustic names of these bones, which remind one of some horny-handed blacksmith in a Thomas Hardy novel, deserve comment. In humans, the stapes does look very like a stirrup. The flat footplate sits in the 'oval window' that's the portal to the inner ear. The footplate is suspended on two separate prongs that join together further up, like a wishbone, or, indeed, a stirrup. The hole between the two prongs is penetrated by a blood vessel, the stapedial artery. Once we have a stirrup, it's only natural to name the other bones the hammer and the anvil, even though they do not look particularly like their ferrous namesakes. The stapes is the smallest bone in the human body; the malleus and incus not much bigger. Together, these bones form the 'ossicles' or 'little bones' of the middle ear.
6. This is so in childhood, at least. Sensitivity to the higher frequencies tends to decline with age, especially in those of us who spent their youth listening to, oh, I don't know, Deep Purple.
7. See H. Heffner, 'Hearing in large and small dogs (*Canis familiaris*)', *Journal of the Acoustical Society of America* **60**, S88, 1976.

8. See R. S. Heffner, 'Primate hearing from a mammalian perspective', *The Anatomical Record* **281A**, 1111–1122, 2004.
9. See K. Ralls, 'Auditory sensitivity in mice: *Peromyscus* and *Mus musculus*', *Animal Behaviour* **15**, 123–128, 1967.
10. R. S. Heffner and H. E. Heffner, 'Hearing range of the domestic cat', *Hearing Research* **19**, 85–88, 1985.
11. See Kastelein *et al.*, 'Audiogram of a striped dolphin (*Stenella coeruleoalba*)', *Journal of the Acoustical Society of America* **113**, 1130, 2003.
12. For a comprehensive recent survey of this remarkable transformation, and much else on the early history of mammals, see Z-X. Luo, 'Transformation and diversification in early mammal evolution', *Nature* **450**, 1011–1019, 2007.
13. See Lautenschlager *et al.*, 'The role of miniaturization in the evolution of the mammalian jaw and middle ear', *Nature* **561**, 533–537, 2018.
14. It almost certainly had whiskers. The fur, however, is conjectural.
15. See Jones *et al.*, 'Regionalization of the axial skeleton predates functional adaptation in the forerunners of mammals', *Nature Ecology & Evolution* **4**, 470–478, 2020.
16. A reconstruction of the ear of *Morganucodon* suggests that it might have been sensitive to sounds as high as 10 kHz. See J. J. Rosowski and A. Graybeal, 'What did *Morganucodon* hear?', *Zoological Journal of the Linnean Society* **101**, 131–168, 2008.
17. See Gill *et al.*, 'Dietary specializations and diversity in feeding ecology of the earliest stem mammals', *Nature* **512**, 303–305, 2014.
18. See E. A. Hoffman and T. B. Rowe, 'Jurassic stem-mammal perinates and the origin of mammalian reproduction', *Nature* **561**, 104–108, 2018.
19. See Hu *et al.*, 'Large Mesozoic mammals fed on young dinosaurs', *Nature* **433**, 149–152, 2005; A. Weil, 'Living large in the Cretaceous', *Nature* **433**, 116–117, 2005.
20. See Meng *et al.*, 'A Mesozoic gliding mammal from northeastern China', *Nature* **444**, 889–893, 2006. This creature, *Volaticotherium*, from the late Jurassic of Inner Mongolia, was

later found to be a member of a group called the triconodonts. These were distinct from the haramiyids, a very ancient mammalian group, which also took to the air; see for example Meng *et al.*, 'New gliding mammaliaforms from the Jurassic', *Nature* **548**, 291–296, 2017; Han *et al.*, 'A Jurassic gliding euharamiyidan mammal with an ear of five auditory bones', *Nature* **551**, 451–456, 2017.

21. See Ji *et al.*, 'A swimming mammaliaform from the Middle Jurassic and ecomorphological diversification of early mammals', *Science* **311**, 1123–1127, 2006.
22. See Krause *et al.*, 'First cranial remains of a gondwanatherian mammal reveal remarkable mosaicism', *Nature* **515**, 512–517, 2014; A. Weil, 'A beast of the southern wild', *Nature* **515**, 495–496, 2014; Krause *et al.*, 'Skeleton of a Cretaceous mammal from Madagascar reflects long-term insularity', *Nature* **581**, 421–427, 2020.
23. See for example Luo *et al.*, 'Dual origin of tribosphenic mammals', *Nature* **409**, 53–57, 2001; A. Weil, 'Relationships to chew over', *Nature* **409**, 28–31, 2001; Rauhut *et al.*, 'A Jurassic mammal from South America', *Nature* **416**, 165–168, 2002.
24. See Bi *et al.*, 'An early Cretaceous eutherian and the placental-marsupial dichotomy', *Nature* **558**, 390–395, 2018; Luo *et al.*, 'A Jurassic eutherian mammal and divergence of marsupials and placentals', *Nature* **476**, 442–445, 2011; Ji *et al.*, 'The earliest known eutherian mammal', *Nature* **416**, 816–822, 2002.
25. See Luo *et al.*, 'An Early Cretaceous tribosphenic mammal and metatherian evolution', *Science* **302**, 1934–1940, 2003.
26. Pantodonts and dinocerates were once grouped together in a single group, the amblypods. When I discovered this as an undergraduate, I was so charmed by the name that I telephoned my mother that very day to inform her of this fact (this was from a telephone box – mobile phones were not widely available). I told her that there was once a group of large, slow-moving herbivores, rather like rhinos, or hippos, and they were called amblypods. 'That's nice, dear,' said my mother, 'you can just imagine them, ambling their pods.'

27. For an excellent guide to mammalian evolution, see D. R. Prothero, *The Princeton Field Guide to Prehistoric Mammals* (Princeton: Princeton University Press, 2017).
28. See Head *et al.*, 'Giant boid snake from the Palaeocene neotropics reveals hotter past equatorial temperatures', *Nature* **457**, 715–717, 2009; M. Huber, 'Snakes tell a torrid tale', *Nature* **457**, 669–671, 2009.
29. See Thewissen *et al.*, 'Skeletons of terrestrial cetaceans and the relationship of whales to artiodactyls', *Nature* **413**, 277–281, 2001; C. de Muizon, 'Walking with Whales', *Nature* **413**, 259–260, 2001.
30. See Thewissen *et al.*, 'Fossil evidence for the origin of aquatic locomotion in archaeocete whales', *Science* **263**, 210–212, 1994.
31. See Gingerich *et al.*, 'Hind limbs of Eocene *Basilosaurus*: evidence of feet in whales', *Science* **249**, 154–157, 1990.
32. For more about the evolution of whales see J. G. M. 'Hans' Thewissen, *The Walking Whales: From Land to Water in Eight Million Years* (Oakland: University of California Press, 2014).
33. See Madsen *et al.*, 'Parallel adaptive radiations in two major clades of placental mammals', *Nature* **409**, 610–614, 2001.

9 PLANET OF THE APES

1. The most primitive primates – the prosimians – include today's lemurs (confined to Madagascar) and a few others, such as bushbabies and tarsiers. The earliest known tarsiers were established 55 million years ago, suggesting that the anthropoids – the group that includes monkeys, apes and humans – was likewise in existence (see Ni *et al.*, 'The oldest known primate skeleton and early haplorhine evolution', *Nature* **498**, 60–63, 2013). The earliest known representatives of the anthropoids, also from the Eocene, were already very diverse, suggesting a long history (see Gebo *et al.*, 'The oldest known anthropoid postcranial fossils and the early evolution of higher primates', *Nature* **404**, 276–278, 2000; Jaeger *et al.*, 'Late middle Eocene epoch of Libya yields earliest known

radiation of African anthropoids', *Nature* **467**, 1095–1098, 2010). Anthropoids had split into monkeys and apes by the Oligocene, at least 25 million years ago (see Stevens *et al.*, 'Oligocene divergence between Old World monkeys and apes', *Nature* **497**, 611–614, 2013).

2. Some tropical grasses exploited a hitherto little-used means of photosynthesis known to biochemists as the 'C4 pathway'. Little used, because it is more elaborate than the 'C3 pathway' used by most plants. The C4 pathway, however, makes more efficient use of carbon dioxide. When carbon dioxide is abundant in the atmosphere, there is little value in using the C4 pathway. But the plants had, perhaps, sensed a long-term change in the Earth's atmosphere; that is, a slow, progressive decrease in the amount of carbon dioxide. See for example C. P. Osborne and L. Sack, 'Evolution of C4 plants: a new hypothesis for an interaction of CO_2 and water relations mediated by plant hydraulics', *Philosophical Transactions of the Royal Society of London* B **367**, 583–600, 2012.
3. De Bonis *et al.*, 'New hominid skull material from the late Miocene of Macedonia in Northern Greece', *Nature* **345**, 712–714, 1990.
4. See Alpagut *et al.*, 'A new specimen of *Ankarapithecus meteai* from the Sinap Formation of central Anatolia', *Nature* **382**, 349–351, 1996.
5. See Suwa *et al.*, 'A new species of great ape from the late Miocene epoch in Ethiopia', *Nature* **448**, 921–924, 2007.
6. See Chaimanee *et al.*, 'A new orang-utan relative from the Late Miocene of Thailand', *Nature* **427**, 439–441, 2004.
7. Perhaps the largest ape ever to have lived was *Gigantopithecus*, which lived in South East Asia in the Pleistocene. It may have been twice the size of a gorilla – though this is hard to estimate as it is known only from teeth and jaw fragments. A study of proteins from tooth enamel show it to have been a relative of orangutans. See Welker *et al.*, 'Enamel proteome shows that *Gigantopithecus* was an early diverging pongine', *Nature* **576**, 262–265, 2019.
8. See Böhme *et al.*, 'A new Miocene ape and locomotion in the ancestor of great apes and humans', *Nature* **575**, 489–493,

2019, with this commentary by Tracy L. Kivell, 'Fossil ape hints and how walking on two feet evolved', *Nature* **575**, 445–446, 2019.

9. See Rook *et al.*, '*Oreopithecus* was a bipedal ape after all: evidence from the iliac cancellous architecture', *Proceedings of the National Academy of Sciences of the United States of America* **96**, 8795–8799, 1999.

10. There were never any apes in the Americas. Apes evolved from Old-World monkeys: the monkeys in the New World are only distant relatives, which may have evolved from immigrants that reached the Americas from Africa in the Eocene (see Bond *et al.*, 'Eocene primates of South America and the African origins of New World monkeys', *Nature* **520**, 538–541, 2015). They are distinguished from their Old-World cousins in retaining long tails, which are often capable of gripping, and function as a fifth limb. This could be a reason why, in the Americas, monkeys remained monkeys, and did not evolve any ape-like or even ground-living forms such as the almost-tailless macaques of the Old World.

11. I should add a note to dispel any confusion between the terms 'hominin' and 'hominid'. The term 'hominid' used to refer to any member of the family Hominidae, which included modern humans and any extinct relatives of humans that weren't more closely related to the Great Apes, or pongids, in the family Pongidae. In recent years it has become clear that the Pongidae does not form a 'natural' group: that is, a group in which all members share the same common ancestor exclusively. It turns out that humans are more closely related to chimps than either is to the gorilla, with the orangutan standing at a more remote remove. This means that the family Pongidae cannot share a common ancestry that does not also include the ancestry of the Hominidae. To resolve this, the definition of the family Hominidae has been expanded to include all the Great Apes as well as humans, and the name hominin (member of the subtribe Hominina of the tribe Hominini of the subfamily Homininae) is used to refer to modern humans and any extinct relatives of humans that aren't more closely related to chimpanzees

– and that's how I use the term here. Matters are muddled still more by conflicting usage. Some researchers now use the term 'hominin' in this sense, whereas others persist in using 'hominid', and some of these two groups have changed their minds with time, making reading some of the literature to which I refer somewhat confusing.

12. See Brunet *et al.*, 'A new hominid from the Upper Miocene of Chad, Central Africa', *Nature* **418**, 145–151, 2002; and Vignaud *et al.*, 'Geology and palaeontology of the Upper Miocene Toros-Menalla hominid locality, Chad', *Nature* **418**, 152–155, 2002. Bernard Wood wrote an accompanying commentary, 'Hominid revelations from Chad', *Nature* **418**, 133–135, 2002.
13. The discoverers of the skull of *Sahelanthropus* named it 'Toumaï'. In Goran, the language of the people who cling on to life in this inhospitable region, this means 'hope of life'.
14. See Haile-Selassie *et al.*, 'Late Miocene hominids from the Middle Awash, Ethiopia', *Nature* **412**, 178–181, 2001.
15. Pickford *et al.*, 'Bipedalism in *Orrorin tugenensis* revealed by its femora', *Comptes Rendus Palevol* **1**, 191–203, 2002.
16. Most discoveries in human evolution dating from 5 million years ago onwards have been made in a narrow strip of Africa extending from Malawi in the south, northwards through Tanzania, Kenya and Ethiopia. This is the great Rift Valley, a slowly widening gash created as two sections of the Earth's crust are being torn apart by the fingernail-fast forces of plate tectonics. Gigantic chunks of rift wall slump into the ever-widening space: the effects of rain and sun erode this into sediment. As the plates pull apart, magma spits and bubbles up from beneath, creating volcanoes. Rivers and lakes are forever forming, merging, expanding and shrinking on the valley floor. The combination of sedimentation, lakes and volcanoes is ideal for fossilization, and it is from the lakeshore sediments of the Rift of Kenya, Tanzania and Ethiopia that the lion's share of evidence for human evolution has been collected. Most of the rest comes from old, eroded limestone caves in a small area of South Africa known

as the 'Cradle of Humankind'. Cave sediments are notoriously hard to date, though there has been some progress. See for example Pickering *et al.*, 'U-Pb-dated flowstones restrict South African early hominin record to dry climate phases', *Nature* **565**, 226–229, 2019. The Earth still moves, and having moved, moves on: in a few million years, Africa east of the Rift will have broken away from its parent continent. The sea will break in to fill the void. The Rift is a new ocean caught in the act of being born; rather like the rift in eastern North America at the end of the Triassic that brought the Atlantic Ocean into being, but without the drama.

17. And one that babies still retain.
18. See Whitcome *et al.*, 'Fetal load and the evolution of lumbar lordosis in bipedal hominins', *Nature* **450**, 1075–1078, 2007.
19. See for example Wilson *et al.*, 'Biomechanics of predator-prey arms race in lion, zebra, cheetah and impala', *Nature* **554**, 183–188, 2018; and the accompanying commentary by Biewener, 'Evolutionary race as predators hunt prey', *Nature* **554**, 176–178, 2018.
20. Other bipedal mammals include kangaroos, and various bounding rodents such as jerboas; but kangaroos support an upright stance with the help of a long tail, and the bounding rodents tend to hop, using both feet at once.
21. Something I discovered for myself when I broke an ankle in a trivial accident at home in August 2018. This mishap left me entirely helpless, a state ameliorated by the instantly accessible ministrations of the almost incomprehensibly complex and vast apparatus that is the National Health Service; including an ambulance, a fully equipped teaching hospital, paramedics, nurses, anaesthetists, surgeons, not to mention an army of support staff, and – when I left hospital – physiotherapists; the loan of a wheelchair from the Red Cross; and (mostly) the care of the long-suffering Mrs Gee, who decided at least partly on the strength of it to enrol for a degree in nursing, specializing in patients with learning disabilities (go figure). The National Health Service is the largest employer not just in Britain but in the whole of

Europe, and consumes a sizeable wedge of Britain's public expenditure. Without such backup, an early hominin that broke its ankle on the African savannah would probably have been killed and eaten.

22. See White *et al.*, '*Australopithecus ramidus*, a new species of early hominid from Aramis, Ethiopia', *Nature* **371**, 306–312, 1994.
23. See A. Gibbons, 'A rare 4.4-million-year-old skeleton has drawn back the curtain of time to reveal the surprising body plan and ecology of our earliest ancestors', *Science* **326**, 1598–1599, 2009.
24. See Leakey *et al.*, 'New four-million-year-old hominid species from Kanapoi and Allia Bay, Kenya', *Nature* **376**, 565–571 (1995); Haile-Selassie *et al.*, 'A 3.8-million-year-old hominin cranium from Woranso-Mille, Ethiopia', *Nature* **573**, 214–219, 2019; F. Spoor, 'Elusive cranium of early hominin found', *Nature* **573**, 200–202, 2019.
25. Johanson *et al.*, 'A new species of the genus *Australopithecus* (Primates, Hominidae) from the Pliocene of Eastern Africa', *Kirtlandia* **28**, 1–14, 1978. At least two other species are known to have lived in the area in the same period. See Haile-Selassie *et al.*, 'New species from Ethiopia further expands Middle Pliocene hominin diversity', *Nature* **521**, 483–488, 2015; F. Spoor, 'The Middle Pliocene gets crowded', *Nature* **521**, 432–433, 2015; Leakey *et al.*, 'New hominin genus from eastern Africa shows diverse middle Pliocene lineages', *Nature* **410**, 433–440, 2001; D. Lieberman, 'Another face in our family tree', *Nature* **410**, 419–420, 2001.
26. Where a very similar creature was named *Australopithecus bahrelghazali*: Brunet *et al.*, 'The first australopithecine 2,500 kilometres west of the Rift Valley (Chad)', *Nature* **378**, 273–275, 1995.
27. As revealed by footprints deposited in wet volcanic ash and preserved at Laetoli in Tanzania. The hominin prints occur in two separate places. In one, a hominin is walking alone. In the other, a hominin appears to be accompanied by a child that is possibly following the adult. See M. D. Leakey and R. L. Hay, 'Pliocene footprints in the Laetolil Beds and Laetoli, northern Tanzania', *Nature* **278**, 317–323, 1979.

28. Having said that, fractures on the most complete specimen, the famous skeleton known as 'Lucy', suggest that she died from wounds sustained by falling out of a tree. See Kappelman *et al.*, 'Perimortem fractures in Lucy suggest mortality from fall out of a tree', *Nature* **537**, 503–507, 2016.
29. See Cerling *et al.*, 'Woody cover and hominin environments in the past 6 million years', *Nature* **476**, 51–56, 2011; C. S. Feibel, 'Shades of the savannah', *Nature* **476**, 39–40, 2011.
30. Haile-Selassie *et al.*, 'A new hominin foot from Ethiopia shows multiple Pliocene bipedal adaptations', *Nature* **483**, 565–569, 2012; D. Lieberman, 'Those feet in ancient times', *Nature* **483**, 550–551, 2012.
31. These included various species of *Australopithecus* and *Homo*, such as *Australopithecus garhi* (see Asfaw *et al.*, '*Australopithecus garhi*: a new species of early hominid from Ethiopia', *Science* **284**, 629–635, 1999); *Australopithecus sediba* (Berger *et al.*, '*Australopithecus sediba*: a new species of Homo-like australopith from South Africa', *Science* **328**, 195–204, 2010); *Homo habilis*, and *Homo rudolfensis* (see Spoor *et al.*, 'Reconstructed *Homo habilis* type OH7 suggests deep-rooted species diversity in early *Homo*', *Nature* **519**, 83–86, 2015); and *Homo naledi* (Berger *et al.*, '*Homo naledi*, a new species of the genus *Homo* from the Dinaledi Chamber, South Africa', *eLife* 2015; 4: e09560). The relationships between all these creatures is a matter of considerable debate. Although the appellation of *Homo* was originally meant to reflect greater brain size and technological capacity (see L. S. B. Leakey, 'A New Fossil Skull from Olduvai', *Nature* **184**, 491–493, 1959; Leakey *et al.*, 'A New Species of the Genus *Homo* from Olduvai Gorge', *Nature* **202**, 7–9, 1964), the discovery of stone tools significantly antedating the earliest *Homo* – around 3.3 million years ago or so – has placed that distinction in doubt. Indeed, a good case has been made that the very earliest species of *Homo* were too little different from *Australopithecus* to merit the distinction: see B. Wood and M. Collard, 'The Human Genus', *Science* **284**, 65–71, 1999.
32. See Harmand *et al.*, '3.3-million-year-old stone tools from Lomekwi 3, West Turkana, Kenya', *Nature* **521**, 310–315,

2015; E. Hovers, 'Tools go back in time', *Nature* **521**, 294–295, 2015; McPherron *et al.*, 'Evidence for stone-tool-assisted consumption of animal tissues before 3.39 million years ago at Dikika, Ethiopia', *Nature* **466**, 857–860, 2010; D. Braun, 'Australopithecine butchers', *Nature* **466**, 828, 2010.

33. The earliest tools were no more sophisticated than the kinds of tools chimpanzees use today, and are very hard to distinguish from rocks chipped by other, natural processes. Indeed, several species of primates, not just hominins, are known to select pebbles and move them to particular areas for use. Some of these artefacts are hard to distinguish from those attributed to early hominins. See Haslam *et al.*, 'Primate archaeology evolves', *Nature Ecology & Evolution* **1**, 1431–1437, 2017.
34. See K. D. Zink and D. E. Lieberman, 'Impact of meat and Lower Palaeolithic food processing techniques on chewing in humans', *Nature* **531**, 500–503, 2016.

10 ACROSS THE WORLD

1. . . . which is the same thing as 23.5 degrees from the vertical, only expressed as a divergence from the horizontal. The two values add up to 90 degrees.
2. And the same, too, for the stars of the southern hemisphere. However, as it happens, the celestial south polar region is an especially dull and boring patch of sky, with nothing much to recommend it, and certainly no prominent stars to mark the south polar equivalent of Polaris.
3. This was worked out by a mathematician called Milutin Milankovic (1879–1958), who did the whole thing without a computer. Imagine.
4. This is one of my very few genuine discoveries, which lies, unread, in my PhD thesis.
5. Apart from the fact that I am British, and happened to study the ice-age fauna of Britain for my PhD thesis, there is a good reason for choosing Britain as an example. As an island

on the western edge of a large landmass, it was prey to the greatest extremes of climate change during this period, and so is a good example of the whole. That's my excuse. And I'm sticking to it.

6. See G. A. Jones, 'A stop-start ocean conveyer', *Nature* **349**, 364–365, 1991.
7. These sudden pulses of iceberg calving are known as Heinrich events. See Bassis *et al.*, 'Heinrich events triggered by ocean forcing and modulated by iostatic adjustment', *Nature* **542**, 332–334, 2017; A. Vieli, 'Pulsating ice sheet', *Nature* **542**, 298–299, 2017.
8. This is captured rather startlingly on the ground. Fossil beds in Ethiopia that bracket the transition show a marked decrease in species fond of mixed woodland, such as *Australopithecus*, and an increase in open country species, such as horses, camels – and *Homo*. See Alemseged *et al.*, 'Fossils from Mille-Logya, Afar, Ethiopia, elucidate the link between Pliocene environmental change and *Homo* origins', *Nature Communications* **11**, 2480 (2020).
9. See D. Bramble and D. Lieberman, 'Endurance running and the evolution of *Homo*', *Nature* **432**, 345–352, 2004, for a compelling essay on the importance of endurance running in the human story. I should add that their exegesis on anatomy concerns *Homo sapiens* rather than *Homo erectus* in particular, so I have taken some liberties. Having said that, *Homo erectus* was the earliest hominin with a body shape closely similar to that of modern humans.
10. The term 'tribe' in this sense refers to a distinct group of individuals bound by kinship and tradition that lives more or less in the same place, and is culturally and more or less genetically distinct from other such groups.
11. A comparison of rates of lethal violence in mammals shows that hominins, and primates, are more violent than mammals in general. See Gómez *et al.*, 'The phylogenetic roots of human lethal violence', *Nature* **538**, 233–237, 2016, with an accompanying commentary by Pagel, 'Lethal violence deep in the human lineage', *Nature* **538**, 180–181, 2016.
12. . . . with a tiny penis. The erect male member in a gorilla

is about 3 centimetres long. Even an average human male can add 10 centimetres to that. See M. Maslin, 'Why did humans evolve big penises but small testicles?' *The Conversation*, 25 January 2017, accessed 1 April 2021; Veale *et al.*, 'Am I normal? A systemic review and construction of nomograms for flaccid and erect penis length and circumference in up to 15,521 men', *BJU International* **115**, 978–986, 2015.

13. See S. Eliassen and C. Jørgensen, 'Extra-pair mating and evolution of cooperative neighbourhoods', *PLoS ONE* doi.org/10.1371./journal.pone.0099878, 2014; B. C. Sheldon and M. Mangel, 'Love thy neighbour', *Nature* **512**, 381–382, 2014.
14. Alan Walker and Pat Shipman describe *Homo erectus* as such in their insightful book *The Wisdom of Bones* (Vintage, 1997).
15. See Dean *et al.*, 'Growth processes in teeth distinguish modern humans from *Homo erectus* and earlier hominins', *Nature* **414**, 628–631, 2001; and the accompanying commentary by Moggi-Cecchi, 'Questions of growth', *Nature* **414**, 595–597, 2001.
16. Although the earliest known Acheulean tools are found in Africa (see for example Asfaw *et al.*, 'The earliest Acheulean from Konso-Gardula', *Nature* **360**, 732–735, 1992), the culture as a whole is named for St-Acheul, an archaeological site in France where it was first recognized.
17. © *The Atlantic Monthly*, 1975.
18. See Joordens *et al.*, '*Homo erectus* at Trinil on Java used shells for tool production and engraving', *Nature* **518**, 228–231, 2015.
19. People have always been surprised to learn that humans are very closely related to chimpanzees, the gorilla, and the orangutan. Religious considerations aside, humans are shockingly different from these creatures. The reason is that just as humans have changed greatly from the common ancestor that we share with apes, the apes have changed much less.
20. The earliest known fossil attributable to *Homo erectus* is a part of a skull from Drimolen Cave in South Africa, dated to just over 2 million years ago – see Herries *et al.*, 'Contemporaneity of *Australopithecus*, *Paranthropus* and early

Homo erectus in South Africa', *Science* **368** doi: 10.1126/science.aaw7293, 2020. The most complete example of African *Homo erectus* is the skeleton of a youth from Kenya: see Brown *et al*., 'Early *Homo erectus* skeleton from west Lake Turkana', Kenya', *Nature* **316**, 788–792, 1985. The long, rangy form of the skeleton is a marked contrast to the more squat frames of earlier hominins.

21. See Zhu *et al*., 'Hominin occupation of the Chinese Loess Plateau since about 2.1 million years ago', *Nature* **559**, 608–612, 2018.
22. See Shen *et al*., 'Age of Zhoukoudian *Homo erectus* determined with 26Al/10Be burial dating', *Nature* **458**, 198–200, 2009; and the accompanying commentary by Ciochon and Bettis, 'Asian *Homo erectus* converges in time', *Nature* **458**, 153–154, 2009.
23. See J. Schwartz, 'Why constrain hominid taxic diversity?', *Nature Ecology & Evolution*, 5 August 2019, https://doi.org/10.1038/s41559-019-0959-2 for a trenchant argument in favour of taxonomic diversity in *Homo erectus*.
24. Whereas all species formally have a binomial name consisting of a genus (*Homo*) and a species (such as *sapiens*), and might also acquire a subspecific name (such as, well, *sapiens*, to give *Homo sapiens sapiens*), these ancient people have lately acquired a tetranomial, *Homo erectus ergaster georgicus*, a handle unique in the annals of nomenclature except perhaps for members of the British royal family, and which only underlines the fact that membership of *Homo erectus* is an extremely broad church. See L. Gabunia and A. Vekua, 'A Plio-Pleistocene hominid from Dmanisi, East Georgia, Caucasus', *Nature* **373**, 509–512, 1995; Lordkipanidze *et al*., 'A complete skull from Dmanisi, Georgia, and the evolutionary biology of early *Homo*', *Science* **342**, 326–331 (2013), for this remarkable name, and a discussion on the very real problems of shoehorning fossil specimens into what might have been species of unknown degrees of variation.
25. See Rizal *et al*., 'Last appearance of *Homo erectus* at Ngandong, Java, 117,000–108,000 years ago', *Nature* **577**, 381–385, 2020.
26. See Swisher *et al*., 'Latest *Homo erectus* of Java: potential

contemporaneity with *Homo sapiens* in Southeast Asia', *Science* **274**, 1870–1874, 1996.

27. See Ingicco *et al.*, 'Earliest known hominin activity in the Philippines by 709 thousand years ago', *Nature* **557**, 233–237, 2018.
28. See Détroit *et al.*, 'A new species of *Homo* from the Late Pleistocene of the Philippines', *Nature* **568**, 181–186, 2019, and the accompanying commentary by Tocheri, 'Previously unknown human species found in Asia raises questions about early hominin dispersals from Africa', *Nature* **568**, 176–178, 2019.
29. See Brown *et al.*, 'A new small-bodied hominin from the Late Pleistocene of Flores, Indonesia', *Nature* **431**, 1055–1061, 2004, with the accompanying commentary by Mirazón Lahr and Foley, 'Human evolution writ small', *Nature* **431**, 1043–1044, 2004; Morwood *et al.*, 'Further evidence for small-bodied hominins from the Late Pleistocene of Flores, Indonesia', *Nature* **437**, 1012–1017, 2005 and the online collection 'The Hobbit at 10', https://www.nature.com/collections/baiecchdeh.
30. See Sutikna *et al.*, 'Revised stratigraphy and chronology for *Homo floresiensis* at Liang Bua in Indonesia', *Nature* **532**, 366–369, 2016; van den Bergh *et al.*, '*Homo floresiensis*-like fossils from the early Middle Pleistocene of Flores', *Nature* **534**, 245–248, 2016; Brumm *et al.*, 'Early stone technology on Flores and its implications for *Homo floresiensis*', *Nature* **441**, 624–628, 2006.
31. These rats still exist, and are accompanied by middle-sized rats and small rats. When I visited the cave at Liang Bua on Flores, where the first specimens of *Homo floresiensis* had been unearthed, I spent a happy day or so helping Dr Hanneke Meijer sort out the hundreds of rat bones into different size classes, along with hundreds of bat bones, and – Hanneke's special interest – the fewer but highly prized bird bones. These bones had been arduously washed out of every gram of sediment dug out of the ground and placed in sacks marked with the precise 3-D location where the sediment had been found. The camp workers had heaved

the heavy sacks down the hill to the rice paddies, sieved out the bones, and brought them back up for us to study. Any excavation must give enormous credit to the back-breaking work of many behind the scenes who make the major discoveries possible, those announced with fanfare in international journals.

32. Victoria Herridge reminded me to make special mention of dwarf elephants. I can't help imagining that elephants and people each got smaller and smaller until they were microscopic, and, effectively invisible, disappeared from view, like the protagonist in *The Incredible Shrinking Man*.
33. See Bermúdez de Castro *et al.*, 'A hominid from the lower Pleistocene of Atapuerca, Spain: possible ancestor to Neandertals and modern humans', *Science* **276**, 1392–1395, 1997; Parfitt *et al.*, 'Early Pleistocene human occupation at the edge of the boreal zone in northwest Europe', *Nature* **466**, 229–233, 2010, and the accompanying commentary by Roberts and Grün, 'Early human northerners', *Nature* **466**, 189–190, 2010; Ashton *et al.*, 'Hominin footprints from Early Pleistocene Deposits at Happisburgh, UK', *PLoS ONE* https://doi.org/10.1371/journal.pone.0088329, 2014.
34. See Welker *et al.*, 'The dental proteome of *Homo antecesso*r', *Nature* **580**, 235–238, 2020.
35. See H. Thieme, 'Lower Palaeolithic hunting spears from Germany', *Nature* **385**, 807–810, 1997.
36. See Roberts *et al.*, 'A hominid tibia from Middle Pleistocene sediments at Boxgrove, UK', *Nature* **369**, 311–313, 1994.
37. See Arsuaga *et al.*, 'Three new human skulls from the Sima de los Huesos Middle Pleistocene site in Sierra de Atapuerca, Spain', *Nature* **362**, 534–537, 1993.
38. Nuclear DNA shows that the Atapuerca individuals were more closely related to Neanderthals than to any other hominin. See Meyer *et al.*, 'Nuclear DNA sequences from the Middle Pleistocene Sima de los Huesos hominins', *Nature* **531**, 504–507, 2016.
39. See Jaubert *et al.*, 'Early Neanderthal constructions deep in Bruniquel Cave in southwestern France', *Nature* **534**, 111–114, 2016; and the accompanying commentary by

Soressi, 'Neanderthals built underground', *Nature* **534**, 43–44, 2016.

40. The Denisovans get their name from the cave in the Altai Mountains of southern Siberia where their remains were first identified. They do not – yet – have a formal zoological name.
41. See Chen *et al.*, 'A late Middle Pleistocene Denisovan mandible from the Tibetan Plateau', *Nature* **569**, 409–412, 2019.
42. If so, they trod very lightly indeed. A mastodon kill site in southern California dated to around 125,000 years ago has, very controversially, been claimed to have been formed by human agency. If so, this was very much earlier than even the most optimistic advocates of an early human occupation of the Americas, which is 30,000 years ago at the outside. See Holen *et al.*, 'A 130,000-year-old archaeological site in southern California, USA', *Nature* **544**, 479–483, 2017.
43. They are known as 'Denisovans', after Denisova Cave in the Altai Mountains of southern Siberia, where their remains were first identified. See Reich *et al.*, 'Genetic history of an archaic hominin group from Denisova Cave in Siberia', *Nature* **468**, 1053–1060, 2010; and the accompanying commentary by Bustamante and Henn, 'Shadows of early migrations', *Nature* **468**, 1044–1045, 2010.

11 THE END OF PREHISTORY

1. See Navarrete *et al.*, 'Energetics and the evolution of human brain size', *Nature* **480**, 91–93, 2011; R. Potts, 'Big brains explained', *Nature* **480**, 43–44, 2011.
2. Natural selection also favoured male preference for more curvaceous female body shapes: see D. W. Yu and G. H. Shepard, Jr, 'Is beauty in the eye of the beholder?', *Nature* **396**, 321–322, 1998.
3. See K. Hawkes, 'Grandmothers and the evolution of human longevity', *American Journal of Human Biology* **15**, 380–400, 2003. Needless to say, the grandmother hypothesis, like

everything else in the evolution of human life history, is controversial, but it seems to make the most sense to me.

4. This explains why men have nipples. Because females have breasts and nipples, males have nipples too, although smaller and non-functional. They also incur a cost: breast cancer occurs in men as well as women, but is rare. Paradoxically, the evolution of female preferences of mate choice maintain otherwise harmful traits in males. See P. Muralidhar, 'Mating preferences of selfish sex chromosomes', *Nature* **570**, 376–379; M. Kirkpatrick, 'Sex chromosomes manipulate mate choice', *Nature* **570**, 311–312, 2019.
5. I am grateful to Simon Conway Morris for this insight.
6. Jared Diamond speculates the rise in Type 2 diabetes, especially among people who until recently lived on subsistence diets, is the result of a sudden switch to Western lifestyles in which starvation is abolished, and eating sugary foods to excess is commonplace. See Diamond, 'The double puzzle of diabetes', *Nature* **423**, 599–602, 2003.
7. *Homo rhodesiensis*, a creature similar to *Homo heidelbergensis*, lived in Central Africa around 300,000 years ago (Grün *et al.*, 'Dating the skull from Broken Hill, Zambia, and its position in human evolution', *Nature* **580**, 372–375, 2020), but there were others. A species of hominin with a remarkably archaic skull lived in Nigeria until as recently as 11,000 years ago (Harvati *et al.*, 'The Later Stone Age calvaria from Iwo Eleru, Nigeria: morphology and chronology', *PLoS ONE* https://doi.org/10.1371/journal.pone.0024024, 2011). There is evidence for further archaic species in Africa, preserved only as fragmentary DNA in modern humans – as so many Cheshire cats, fading from view until only their smiles are left (see for example Hsieh *et al.*, 'Model-based analyses of whole-genome data reveal a complex evolutionary history involving archaic introgression in Central African Pygmies', *Genome Research* **26**, 291–300, 2016).
8. The earliest known evidence for the first stirrings of *Homo sapiens* are around 315,000 years old and come from Morocco (see Hublin *et al.*, 'New fossils from Jebel Irhoud, Morocco, and the pan-African origin of *Homo sapiens*', *Nature* **546**,

289–292, 2017; Richter *et al.*, 'The age of the hominin fossils from Jebel Irhoud, Morocco, and the origins of the Middle Stone Age', *Nature* **546**, 293–296, 2017; Stringer and Galway-Witham, 'On the origin of our species', *Nature* **546**, 212–214, 2017). Other early specimens of *Homo sapiens* include remains from Kibish, Ethiopia, dated at around 195,000 years old (McDougall *et al.*, 'Stratigraphic placement and age of modern humans from Kibish, Ethiopia', *Nature* **433**, 733–736, 2005) and the Middle Awash, also in Ethiopia (see White *et al.*, 'Pleistocene *Homo sapiens* from Middle Awash, Ethiopia', *Nature* **423**, 742–747, 2003; Stringer, 'Out of Ethiopia', *Nature* **423**, 693–695, 2003).

9. Harvati *et al.*, 'Apidima Cave fossils provide earliest evidence of *Homo sapiens* in Eurasia', *Nature* **571**, 500–504, 2019; McDermott *et al.*, 'Mass-spectrometric U-series dates for Israeli Neanderthal/early modern hominid sites', *Nature* **363**, 252–255, 1993; Hershkovitz *et al.*, 'The earliest modern humans outside Africa', *Science* **359**, 456–459, 2018.
10. See Chan *et al.*, 'Human origins in a southern African palaeo-wetland and first migrations', *Nature* **575**, 185–189, 2019.
11. See Henshilwood *et al.*, 'A 100,000-year-old Ochre-Processing Workshop at Blombos Cave, South Africa', *Science* **334**, 219–222, 2011.
12. See Henshilwood *et al.*, 'An abstract drawing from the 73,000-year-old levels at Blombos Cave, South Africa', *Nature* **562**, 115–118, 2018.
13. See Brown *et al.*, 'An early and enduring advanced technology originating 71,000 years ago in South Africa', *Nature* **491**, 590–593.
14. See Rito *et al.*, 'A dispersal of *Homo sapiens* from southern to eastern Africa immediately preceded the out-of-Africa migration', *Scientific Reports* **9**, 4728, 2019.
15. Toba dwarfed the famous eruption of Tambora, also in Indonesia, in 1815. That event ushered in 'The Year Without Summer'; when a group of radicals hoping to enjoy a summer holiday were holed up in a villa on Lake Geneva instead, and amused themselves by composing horror stories. One of the company was the teenage Mary Shelley, who came

up with a bodice-ripper called *Frankenstein; or, The Modern Prometheus*. Clearly, something saved for a rainy day.

16. See Smith *et al.*, 'Humans thrived in South Africa through the Toba eruption about 74,000 years ago', *Nature* **555**, 511–515, 2018.
17. See Petraglia *et al.*, 'Middle Paleolithic assemblages from the Indian Subcontinent before and after the Toba super-eruption', *Science* **317**, 114–116, 2007.
18. See Westaway *et al.*, 'An early modern human presence in Sumatra 73,000–63,000 years ago', *Nature* **548**, 322–325, 2017.
19. This has been shown to be the case for australopiths. Chemical analysis of trace elements in the enamel of australopith teeth shows that the smaller individuals – presumed to have been female – moved around during their lives further than males. See Copeland *et al.*, 'Strontium isotope evidence for landscape use by early hominins', *Nature* **474**, 76–78, 2011; M. J. Schoeninger, 'In search of the australopithecines', *Nature* **474**, 43–45, 2011.
20. See A. Timmermann and T. Friedrich, 'Late Pleistocene climate drivers of early human migration'. *Nature* **538**, 92–95, 2016.
21. Clarkson *et al.*, 'Human occupation of northern Australia by 65,000 years ago', *Nature* **547**, 306–310, 2017.
22. See, for example, F. A. Villanea and J. G. Schraiber, 'Multiple episodes of interbreeding between Neanderthals and modern humans', *Nature Ecology & Evolution* **3**, 39–44, 2019, with the accompanying commentary by F. Mafessoni, 'Encounters with archaic hominins', *Nature Ecology & Evolution* **3**, 14–15, 2019; Sankararaman *et al.*, 'The genomic landscape of Neanderthal ancestry in present-day humans', *Nature* **507**, 354–357, 2014.
23. See Huerta-Sánchez *et al.*, 'Altitude adaptation in Tibetans caused by introgression of Denisovan-like DNA', *Nature* **512**, 194–197, 2014.
24. See Hublin *et al.*, 'Initial Upper Palaeolithic *Homo sapiens* from Bacho Kiro Cave, Bulgaria', *Nature* **581**, 299–302, 2020, with the accompanying report by Fewlass *et al.*, 'A 14C chronology for the Middle to Upper Palaeolithic transition

at Bacho Kiro Cave, Bulgaria', *Nature Ecology & Evolution* **4**, 794–801, 2020, and the accompanying commentary by Banks, 'Puzzling out the Middle-to-Upper Palaeolithic transition', *Nature Ecology & Evolution* **4**, 775–776, 2020. See also M. Cortés-Sanchéz *et al.*, 'An early Aurignacian arrival in south-western Europe', *Nature Ecology & Evolution* **3**, 207–212, 2019; Benazzi *et al.*, 'Early dispersal of modern humans in Europe and implications for Neanderthal behaviour', *Nature* **479**, 525–528, 2011.

25. See Higham *et al.*, 'The timing and spatiotemporal patterning of Neanderthal disappearance', *Nature* **512**, 306–309, 2014, and the accompanying commentary by W. Davies, 'The time of the last Neanderthals', *Nature* **512**, 260–261, 2014.
26. 'Are you telling me that they *copulated*?' asked an incredulous elderly member of the audience, in cut-glass tones, of a speaker addressing this sensitive topic at a meeting on ancient DNA at the Royal Society in London. Seated somewhere at the back, I was tempted to stand up and reply, in a similarly imperious tone, that 'not only did they copulate, but their union was blessed with issue!' I stayed in my seat.
27. See Koldony and Feldman, 'A parsimonious neutral model suggests Neanderthal replacement was determined by migration and random species drift', *Nature Communications* **8**, 1040, 2017; and C. Stringer and C. Gamble, *In Search of the Neanderthals* (London: Thames & Hudson, 1994). Similar mechanisms have been observed in other species. The North American grey squirrel, for example, was introduced into England in the eighteenth century. Two hundred years later it had virtually replaced the native red squirrel, by virtue of faster breeding and a more aggressive attitude towards holding territory. See Okubo *et al.*, 'On the spatial spread of the grey squirrel in Britain', *Proceedings of the Royal Society of London* B, **238**, 113–125, 1989.
28. See Zilhão *et al.*, 'Precise dating of the Middle-to-Upper Paleolithic transition in Murcia (Spain) supports late Neandertal persistence in Iberia', *Heliyon* **3**, e00435, 2017.
29. Slimak *et al.*, 'Late Mousterian persistence near the Arctic Circle', *Science* **332**, 841–845, 2011.

30. Vaesen *et al.*, 'Inbreeding, Allee effects and stochasticity might be sufficient to account for Neanderthal extinction', *PLoS ONE* **14**, e0225117, 2019.
31. J. Diamond, 'The last people alive', *Nature* **370**, 331–332, 1994.
32. Fu *et al.*, 'An early modern human from Romania with a recent Neanderthal ancestor', *Nature* **524**, 216–219.
33. Conard *et al.*, 'New flutes document the earliest musical tradition in southwestern Germany', *Nature* **460**, 737–740, 2009.
34. Conard, 'Palaeolithic ivory sculptures from southwestern Germany and the origins of figurative art', *Nature* **426**, 830–832, 2003.
35. See Aubert *et al.*, 'Pleistocene cave art from Sulawesi, Indonesia', *Nature* **514**, 223–227, 2014; Aubert *et al.*, 'Palaeolithic cave art in Borneo', *Nature* **564**, 254–257, 2018.
36. Lubman, 'Did Paleolithic cave artists intentionally paint at resonant cave locations?', *Journal of the Acoustical Society of America*, **141**, 3999, 2017.

12 THE PAST OF THE FUTURE

1. I call this 'The Karenina Principle'. You're welcome.
2. *Dark Eden*, a novel by Chris Beckett (Corvus, 2012) concerns one John Redlantern, one of 532 descendants of two astronauts stranded on a distant planet. It's a poignant tale of a small community's desperate efforts to survive despite the effects of congenital malformation brought on by inbreeding.
3. One thinks of the tragic story of *Dedeckera eurekensis*, a shrub confined to the Mojave Desert. It evolved in milder circumstances, but failure to adapt resulted in a slew of genetic abnormalities that has ensured an almost total failure to reproduce. See Wiens *et al.*, 'Developmental failure and loss of reproductive capacity in the rare palaeoendemic shrub *Dedeckera eurekensis*', *Nature* **338**, 65–67, 1989.
4. See A. Sang *et al.*, 'Indirect evidence for an extinction debt of grassland butterflies half century after habitat loss', *Biological Conservation* **143**, 1405–1413, 2010.

5. See Tilman *et al.*, 'Habitat destruction and the extinction debt', *Nature* **371**, 65–66, 1994.
6. See A. J. Stuart, *Vanished Giants* (Chicago: University of Chicago Press, 2020) for a comprehensive and readable account of the end-Pleistocene extinctions.
7. See Stuart *et al.*, 'Pleistocene to Holocene extinction dynamics in giant deer and woolly mammoth', *Nature* **431**, 684–689, 2004.
8. For example, in my unpublished and unread PhD thesis, *Bovidae from the Pleistocene of Britain* (Fitzwilliam College, University of Cambridge, 1991), I show that a small, rugged kind of bison was common in Britain during the middle of the most recent Cold Stage, but was replaced by a larger form as the Cold Stage progressed. Bison were common during the preceding Ipswichian interglacial, too, but were of a larger sort, and lived in England outside the Thames Valley – in those days, London was aurochs country. In the Hoxnian, an interglacial or two before that, aurochs were common, and nary a bison could be had anywhere, not even for ready money. And even before *that*, in the Cromerian, there were no aurochs, but there were bison – of yet another sort. But Pleistocene sediments in Britain are common and (relatively) easy to arrange in order. Such resolution wouldn't be possible with deposits of, say, Permian age.
9. It's long been thought that human arrival in the Americas could not have been earlier than around 15,000 years ago. However, new archaeology and revised dating methods show that humans were present, if sparsely, some 30,000 years ago, or even earlier. See L. Becerra-Valdivia and T. Higham, 'The timing and effect of the earliest human arrivals in North America', doi.org/10.1038/s41586-020-2491-6, 2020; Ardelean *et al.*, 'Evidence for human occupation in Mexico around the Last Glacial Maximum', *Nature* **584**, 87–92, 2020.
10. The Moon would, too. But as this tale is about life on Earth, this arguably goes beyond my remit.
11. See Piperno *et al.*, 'Processing of wild cereal grains in the Upper Palaeolithic revealed by starch grain analysis', *Nature* **430**, 670–673, 2004.

12. See J. Diamond, 'Evolution, consequences and future of plant and animal domestication', *Nature* **418**, 700–707, 2002.
13. See Krausmann *et al.*, 'Global human appropriation of net primary production doubled in the 20th century', *Proceedings of the National Academy of Sciences of the United States of America* **110**, 10324–10329, 2013.
14. If you want to know, I was born in 1962. 'Good Luck Charm' by Elvis Presley was top of the Billboard Hot 100, also top of the pops in the UK.
15. The Total Fertility Rate (TFR) – the rate at which babies must be born to outpace the death rate – is 2.1 children per mother: it would be 2.0, but a little bit is added on to compensate for early mishap, and the fact that male children are more likely to die than female ones. By 2100, 183 countries (out of 195 studied) will have a TFR lower than this, and the global population will be smaller than it is now. In some countries, such as Spain, Thailand and Japan, the population will have declined by a half by that date. See Vollset *et al.*, 'Fertility, mortality, migration and population scenarios for 195 countries and territories from 2017 to 2100: a forecasting analysis for the Global Burden of Disease Study', *The Lancet* doi.org/10.1016/S0140-6736(20)20677-2, 2020.
16. See Kaessmann *et al.*, 'Great ape DNA sequences reveal a reduced diversity and an expansion in humans', *Nature Genetics* **27**, 155–156, 2001; Kaessmann *et al.*, 'Extensive nuclear DNA sequence diversity among chimpanzees', *Science* **286**, 1159–1162, 1999.
17. I should say that from this point onwards, most of what I say is conjecture, or what scientists call Making Stuff Up. As someone once said, prediction is very hard, especially about the future.
18. I have borrowed this arresting image from *After Man: A Zoology of the Future* (Granada Publishing, 1982), in which Dougal Dixon speculates on the animals that might have evolved 50 million years after the demise of humanity. The 'nightstalker' is a horrific bat-derived carnivore that prowls the nighted forests of a newly formed volcanic landmass

called Batavia, colonized only by bats. The creatures evolve to occupy many un-batlike ecological niches.

19. If you want to lie awake at night worrying, read *The Life and Death of Planet Earth* by Peter Ward and Donald Brownlee (Times Books, Henry Holt and Co., 2002) in which these two factors are remorselessly explored.
20. The atmospheric concentration of carbon dioxide over the past 800,000 years or so has never exceeded about 300 ppm. In 2018, it exceeded 400 ppm, as a result of human activity, a concentration not seen for more than 3 million years. See K. Hashimoto, 'Global temperature and atmospheric carbon dioxide concentration', in *Global Carbon Dioxide Recycling*, SpringerBriefs in Energy (Singapore: Springer, 2019).
21. There is more to it, of course. The picture I have just painted is based on the idea that it is only bare, lifeless silicate rock that is weathered. Although that was true billions of years ago, the presence of life changes the game. The presence of organic matter, and carbonate-rich sedimentary rock, influences the rate of weathering both upwards and downwards, in ways that are hard to predict (R. G. Hilton and A. J. West, 'Mountains, erosion and the carbon cycle', *Nature Reviews Earth & Environment* **1**, 284–299, 2020). In addition, most carbon on land is stored in an entirely life-generated substrate; that is, soil. Increasing temperature stimulates greater respiration in soil microbes, the result of which is a release of carbon dioxide into the atmosphere (Crowther *et al.*, 'Quantifying global soil carbon losses in response to warming', *Nature* **540**, 104–108, 2016). These and other processes influence the transfer of carbon dioxide from the atmosphere to the deep sea.
22. Another complication is that the Earth might have been struck one or more times by asteroids around 800 million years ago: a survey of cratering on the Moon shows an increase in impacts around that time. See Terada *et al.*, 'Asteroid shower on the Earth-Moon system immediately before the Cryogenian period revealed by KAGUYA', *Nature Communications* 11, 3453, 2020.

23. See Simon *et al.*, 'Origin and diversification of endomycorrhizal fungi and coincidence with vascular land plants', *Nature* **363**, 67–69, 1993.
24. See Simard *et al.*, 'Net transfer of carbon between ectomycorrhizal tree species in the field', *Nature* **388**, 579–582, 1997; Song *et al.*, 'Defoliation of interior Douglas-fir elicits carbon transfer and stress signalling to ponderosa pine neighbors through ectomycorrhizal networks', *Scientific Reports* **5**, 8495, 2015; J. Whitfield, 'Underground networking', *Nature* **449**, 136–138, 2007.
25. Smith *et al.*, 'The fungus *Armillaria bulbosa* is among the largest and oldest living organisms', *Nature* **356**, 428–431, 1992.
26. Hymenoptera started to diversify around 281 million years ago (Peters *et al.*, 'Evolutionary history of the Hymenoptera', *Current Biology* **27**, 1013–1018, 2017); the earliest known moths lived 300 million years ago (Kawahara *et al.*, 'Phylogenomics reveals the evolutionary timing and pattern of butterflies and moths', *Proceedings of the National Academy of Sciences of the United States of America* **116**, 22657–22663, 2019).
27. For a useful primer, which explains why, when we eat a fig, we don't get a mouthful of wasps, see J. M. Cook and S. A. West, 'Figs and fig wasps', *Current Biology* **15**, R978–R980, 2005.
28. See C. A. Sheppard and R. A. Oliver, 'Yucca moths and yucca plants: discovery of "the most wonderful case of fertilisation"', *American Entomologist* **50**, 32–46, 2004.
29. See D. M. Gordon, 'The rewards of restraint in the collective regulation of foraging by harvester ant colonies', *Nature* **498**, 91–93, 2013.
30. A theme discussed in book form by E. O. Wilson in *The Social Conquest of Earth* (New York: Liveright, 2012).
31. Scientists are unanimous that there will be a supercontinent in another 250 million years, but opinions differ on its precise shape. One model has it that the Americas will push westwards until they meet eastern Asia, extinguishing the Pacific Ocean. Another holds that the Americas will, as they have done in the past, be drawn towards the western edge of

Eurasia, closing the Atlantic. Ted Nield's book *Supercontinent* explains the reasoning behind these scenarios.

32. For a good introduction to the deep biosphere, see A. L. Mascarelli, 'Low life', *Nature* **459**, 770–773, 2009.
33. See Borgonie *et al.*, 'Eukaryotic opportunists dominate the deep-subsurface biosphere in South Africa', *Nature Communications* **6**, 8952, 2015; Borgonie *et al.*, 'Nematoda from the terrestrial deep subsurface of South Africa', *Nature* **474**, 79–82, 2011.
34. The scientist was one N. A. Cobb, who drew this pen portrait of roundworms in 'Nematodes and their relationships', *United States Department of Agriculture Yearbook* (Washington DC: US Department of Agriculture, 1914), p. 472.
35. Models of the carbon cycle suggest that life will die out between 900 million and 1.5 billion years into the future. A billion years after that, the oceans will boil away. See K. Caldeira and J. F. Kasting, 'The life span of the biosphere revisited', *Nature* **360**, 721–723, 1992. What happens after that depends on how fast the oceans boil. Fast, and the Earth will dry out and become a hot, desert planet. Slow, and much of the atmosphere will shroud the Earth, creating a greenhouse effect so powerful that the surface of the planet will melt. These delightful visions are spelled out by P. Ward and D. Brownlee in *The Life and Death of Planet Earth* (Times Books, Henry Holt and Co., 2002). In the end it will hardly matter: in several more billion years, the Sun will expand into a 'Red Giant' that will fill the sky, frying the Earth to a cinder and possibly consuming it, before shedding most of its mass as a so-called 'planetary nebula' and shrinking to a tiny white-dwarf star that could last for trillions of years. The Sun, massive though it is, is not massive enough to explode and become a supernova, seeding new generations of stars, planets and life.

EPILOGUE

1. See Barnosky *et al.*, 'Has the Earth's sixth mass extinction already arrived?' *Nature* **471**, 51–57, 2011.
2. See https://www.carbonbrief.org/analysis-uk-renewables-generate-more-electricity-than-fossil-fuels-for-first-time, accessed 26 July 2020.
3. See, for example, Paul Ehrlich's book *The Population Bomb*. For an appraisal of its effects half a century on, see https://www.smithsonianmag.com/innovation/book-incited-world-wide-fear-overpopulation-180967499/ – accessed 26 July 2020.
4. See https://ourworldindata.org/energy, accessed 26 July 2020.
5. See Friedman *et al.*, 'Measuring and forecasting progress towards the education-related SDG targets', *Nature* **580**, 636–639, 2020.
6. See Vollset *et al.*, 'Fertility, mortality, migration and population scenarios for 195 countries and territories from 2017 to 2100: a forecasting analysis for the Global Burden of Disease Study', *The Lancet* doi.org/10.1016/S0140-6736(20)20677-2, 2020.
7. See for example Horneck *et al.*, 'Space microbiology', *Microbiology and Molecular Biology Reviews* **74**, 121–156, 2010. The possibility that living things (apart from humans) might be able to travel between planets is something I have chosen not to discuss in this book.
8. . . . and all of them male, which limits the reproductive opportunities somewhat.

Index